MATH AMERICA

Coast-to-Coast Enrichment Activities for Intermediate Math

by
Betty Bobo
and
Lynn Embry

illustrated by Janet Skiles

Cover by Janet Skiles

ISBN No. 0-86653-378-8

Good Apple, Inc.
A Division of Frank Schaffer Publications, Inc.
23740 Hawthorne Boulevard
Torrance, CA 90505-5927

TABLE OF CONTENTS

OBJECTIVES

1. To enrich the intermediate level math curriculum
2. To provide geographic knowledge of the United States
3. To develop an appreciation of the beauty, diversity, and wealth of the United States
4. To create a desire for independent research
5. To provide practice in following directions

MATH AMERICA

Hello boys and girls,

The United States of America is a beautiful country. I love meeting new friends as I travel from coast to coast. I would like to share some interesting information with you through riddles, map mysteries, and decoders.

Our land is full of priceless treasures awaiting your discovery!

Happy calculations!
Professor Ben Around

Can you crack the code to discover a clue for being successful in math?

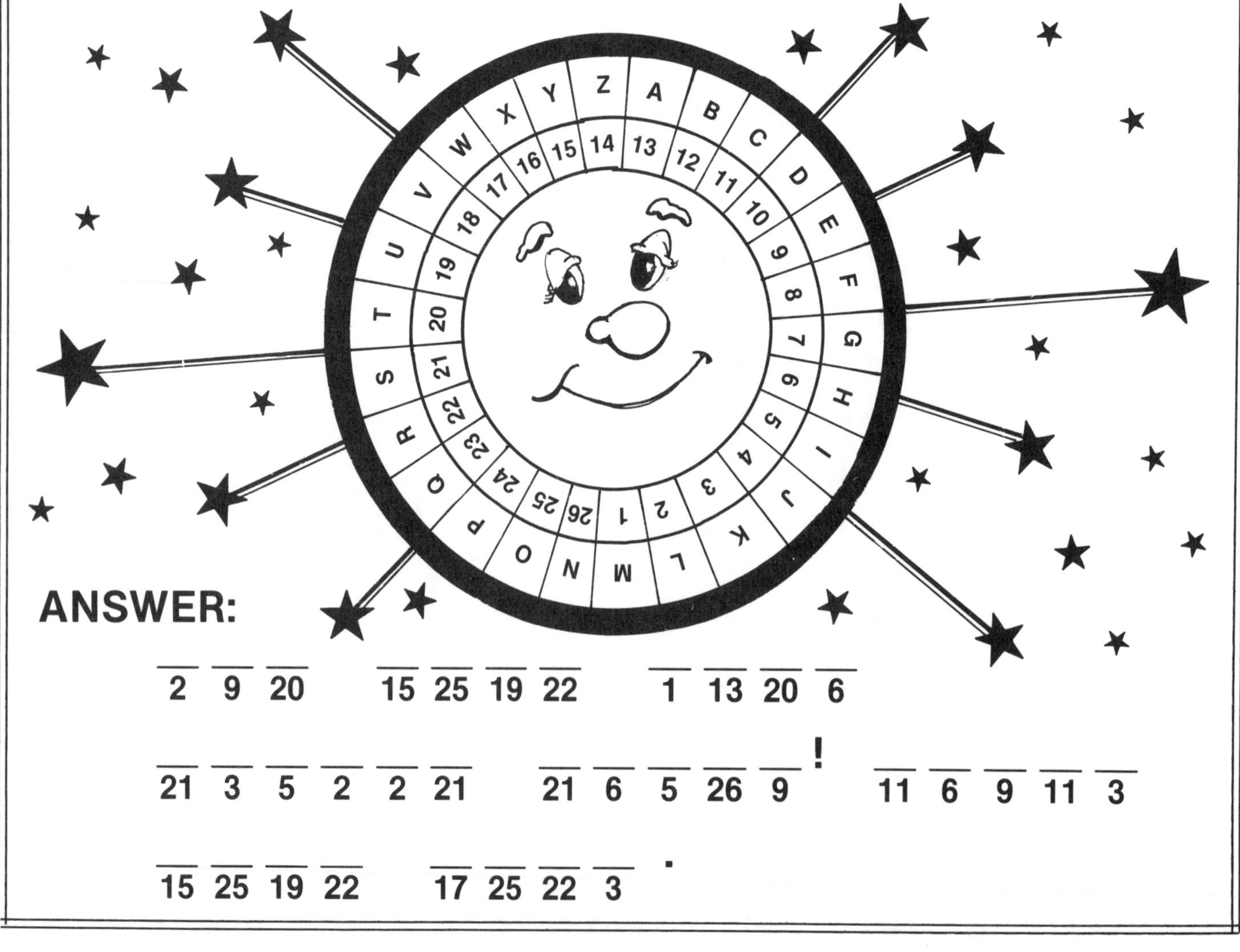

ANSWER:

__ __ __   __ __ __ __   __ __ __ __
2 9 20   15 25 19 22   1 13 20 6

__ __ __ __ __ __   __ __ __ __ __ !   __ __ __ __ __
21 3 5 2 2 21   21 6 5 26 9 !   11 6 9 11 3

__ __ __ __   __ __ __ __ .
15 25 19 22   17 25 22 3 .

NAME ____________________

Hello friends,

Wish you were here with me! I am visiting the southernmost state in the United States. It has only one large city. While lying on black sand beaches, I can see the snowcapped peaks. Can you name this world famous state?

Professor Ben Around

To find out, add. Circle the letter above each odd sum. Unscramble the circled letters to spell the state's name. Write the letters in the blanks below.

C 7 + 1 = ____	A 2 + 9 = ____	N 3 + 1 = ____	I 8 + 9 = ____
W 7 + 6 = ____	Q 9 + 5 = ____	Y 5 + 7 = ____	H 6 + 9 = ____
E 12 + 8 = ____	K 12 + 12 = ____	P 9 + 1 = ____	G 24 + 4 = ____
A 5 + 6 = ____	F 5 + 1 = ____	J 10 + 20 = ____	D 15 + 7 = ____
S 8 + 8 = ____	I 11 + 6 = ____	U 9 + 9 = ____	T 36 + 8 = ____
B 33 + 3 = ____	M 42 + 4 = ____	O 26 + 8 = ____	L 12 + 14 = ____

ANSWER: __ __ __ __ __ __

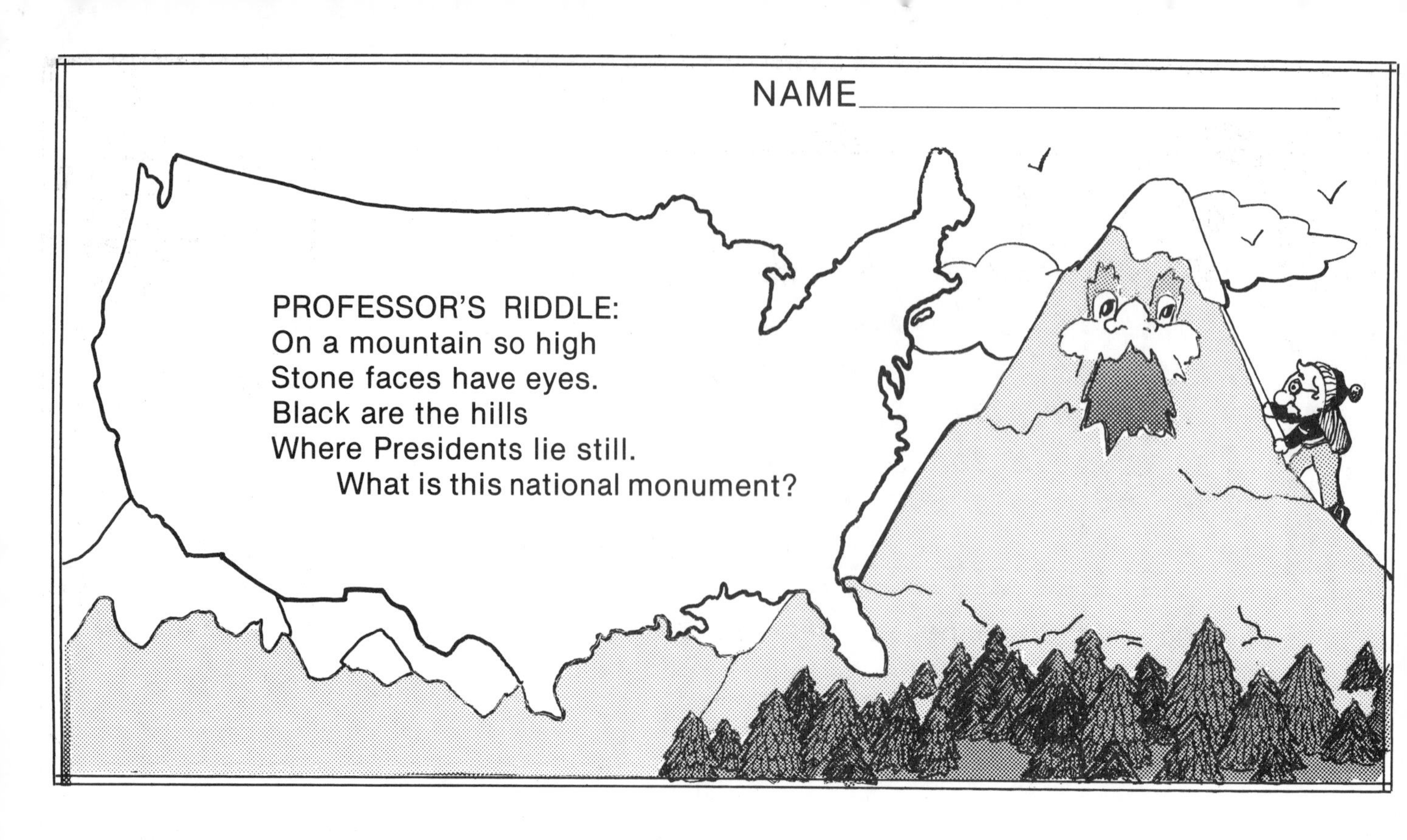

To find out, add. Write the matching letter in the blank above the sum. The letters spell the answer. Some letters will not be used.

ANSWER: ___ ___ ___ ___ ___ ___ ___ ___ ___ ___ ___ ___ ___

21 30 17 25 18 22 17 29 19 21 30 22 24

H 8 + 5 + 6 = ____	U 6 + 3 + 8 = ____	T 4 + 8 + 6 = ____	K 9 + 3 + 8 = ____
D 8 + 6 + 9 = ____	R 6 + 9 + 7 = ____	M 5 + 9 + 7 = ____	S 14 + 6 + 9 = ____
B 13 + 6 + 7 = ____	F 9 + 4 + 3 = ____	O 19 + 4 + 7 = ____	N 11 + 7 + 7 = ____
E 7 + 8 + 9 = ____	R 15 + 5 + 2 = ____	U 3 + 5 + 9 = ____	G 52 + 8 + 2 = ____
M 4 + 9 + 8 = ____	C 25 + 5 + 5 = ____	O 21 + 6 + 3 = ____	V 25 + 4 + 3 = ____

NAME________________________

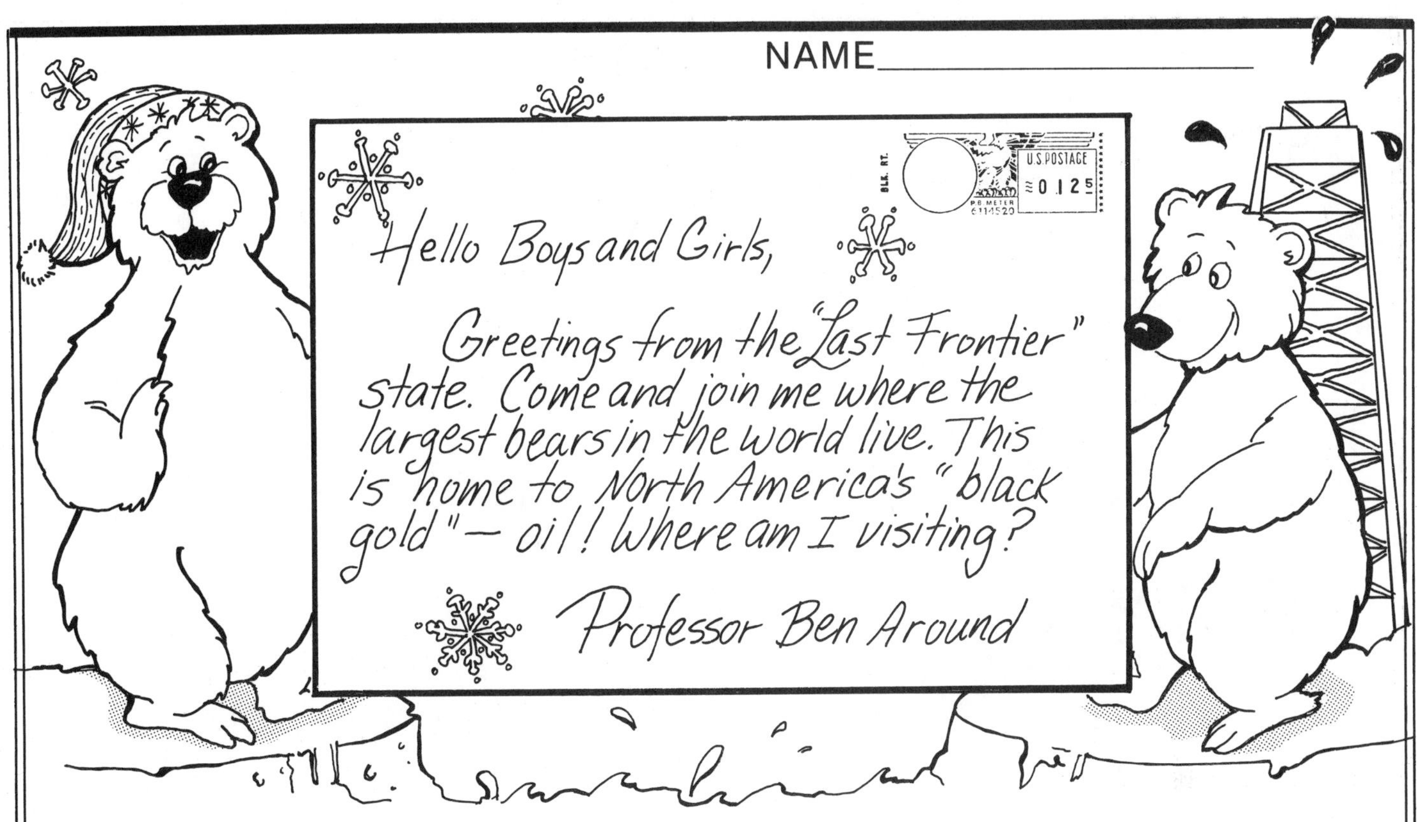

Hello Boys and Girls,

Greetings from the "Last Frontier" state. Come and join me where the largest bears in the world live. This is home to North America's "black gold" — oil! Where am I visiting?

Professor Ben Around

To find out, add. Cross out all sums that are even numbers. Unscramble the letters beside the remaining sums to spell the answer. Write the letters in the blanks below.

85 24 + 31 I	28 16 + 37 L	27 85 + 94 D	55 63 + 43 A
12 89 + 13 H	58 21 + 15 K	58 41 + 23 O	81 76 + 48 S
26 52 + 95 A	75 22 + 45 N	43 35 + 19 K	62 32 + 14 W
12 14 + 18 T	80 45 + 17 R	92 12 + 18 Y	21 59 + 47 A

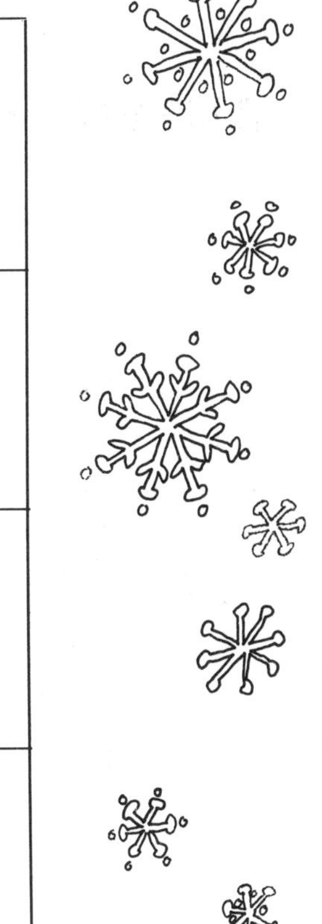

ANSWER: __ __ __ __ __ __ __

NAME________________________

PROFESSOR'S MAP MYSTERY:
Rivers flow to the east and the west
Dividing the land at the Rocky's crest.
Rising high into the sky,
Majestic beauty to the eyes.
What physical feature am I?

To find out, add. Write the matching letter in the square to the right of each problem. The letters spell the answer. Write the letters in the blanks below.

D	O	I	L	T	E	C	A	V	N
70	81	62	68	73	59	75	58	91	72

1. 26 + 49 = ______ 1. [ ]
2. 41 + 40 = ______ 2. [ ]
3. 36 + 36 = ______ 3. [ ]
4. 18 + 55 = ______ 4. [ ]
5. 35 + 27 = ______ 5. [ ]
6. 33 + 39 = ______ 6. [ ]
7. 31 + 28 = ______ 7. [ ]
8. 46 + 26 = ______ 8. [ ]
9. 42 + 31 = ______ 9. [ ]
10. 46 + 12 = ______ 10. [ ]
11. 21 + 47 = ______ 11. [ ]

12. 49 + 21 = ______ 12. [ ]
13. 24 + 38 = ______ 13. [ ]
14. 75 + 16 = ______ 14. [ ]
15. 36 + 26 = ______ 15. [ ]
16. 33 + 37 = ______ 16. [ ]
17. 25 + 34 = ______ 17. [ ]

ANSWER:

_ _ _ _ _ _ _ _ _ _ _   _ _ _ _ _ _

NAME
PROFESSOR'S RIDDLE:
The Colorado runs this way;
Arizona's grandest wonder they say.
Mules journey every day
Down the hills, knowing the way.
What is this wonder of nature?
To find out, add. Write the matching letter in the blank above the sum. The letters spell the answer.
26 + 7 + 8 = ___
A
49 +8 + 3 = ___
D
59 + 3 + 4 = ___
C
65 + 8 + 9 = ___
N
37 + 2 + 3 = ___
G
25 + 6 + 3 = ___
N
46 + 2 + 9 = ___
A
53 + 9 + 5 = ___
N
48 + 7 + 9 = ___
R
19 + 9 + 9 = ___
O
85 + 8 + 2 = ___
Y
ANSWER:
42 64 41 34 60
66 57 67 95 37 82

NAME________________

Greetings,

I'm visiting a capital city located in a big oil field! Outside this state capitol building oil rigs stand. While laws are made, oil is produced for world trade. Do you know what city I'm visiting?

Professor Ben Around

To find out, add. Cross out the letter beside each sum in the decoder. The remaining letters spell the answer. Write the letters in the blanks below.

1. 34 + 58 + 47 = ______

2. 63 + 38 + 26 = ______

3. 74 + 56 + 23 = ______

4. 28 + 16 + 35 = ______

5. 45 + 37 + 25 = ______

6. 24 + 73 + 28 = ______

7. 23 + 41 + 26 = ______

8. 35 + 95 + 48 = ______

9. 24 + 56 + 87 = ______

10. 81 + 65 + 79 = ______

11. 43 + 25 + 78 = ______

12. 72 + 59 + 28 = ______

13. 82 + 76 + 29 = ______

14. 35 + 45 + 48 = ______

15. 23 + 83 + 69 = ______

16. 42 + 58 + 23 = ______

DECODER	
N	79
E	159
O	21
F	146
K	145
B	225
L	218
Q	127
A	181
R	167
H	209
P	107
O	357
X	139
D	125
M	291
G	178
A	148
J	187
C	73
Z	128
I	138
S	153
T	173
Y	85
V	90
O	160
W	175
K	256
U	123

ANSWER: _ _ _ _ _ _ _ _ _ _ _ _ _ _ _ _ _ _, _ _ _

NAME________________________

To find out, add. Write the matching letter in the blank above the sum. The letters spell the answer.

ANSWER:

__ 92 __ 60 __ 85 __ 123 __ 101 __ 90 __ 71 __ 96 __ 57 __ 83 __ 108

__ 97 __ 109 __ 114 __ 117 __ 74 __ 65 __ 67 __ 131   __ 104 __ 81 __ 79 __ 91

NAME ____________________________

PROFESSOR'S MAP MYSTERY:
This roof is 208 feet tall;
Here they like all kinds of ball.
Temperatures stay the same all year,
So people can holler, applaud and cheer.
Can you identify this place?

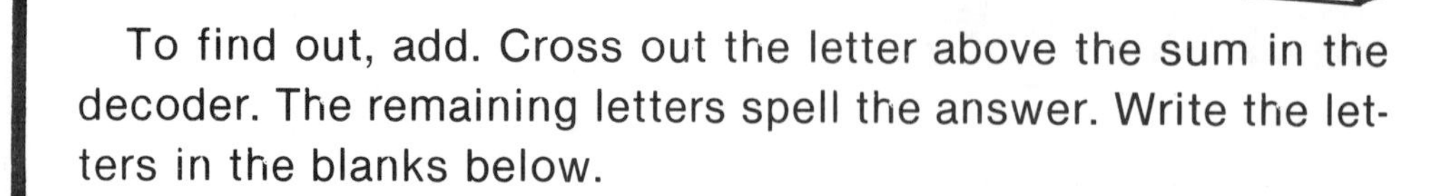

To find out, add. Cross out the letter above the sum in the decoder. The remaining letters spell the answer. Write the letters in the blanks below.

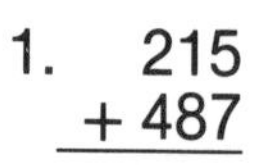

1. 215 + 487
2. 556 + 275
3. 414 + 209
4. 836 + 158
5. 125 + 259
6. 125 + 323
7. 591 + 287
8. 275 + 366
9. 508 + 308
10. 142 + 829
11. 456 + 286
12. 234 + 555
13. 135 + 265
14. 403 + 188
15. 153 + 693

DECODER

H 789	P 384	A 555	Q 400	J 994
C 971	S 429	G 448	N 742	K 816
L 878	F 591	T 673	U 623	I 846
R 981	V 831	O 226	W 641	D 337
O 164	B 702	M 352	E 250	

ANSWER: ___ ___ ___ ___ ___ ___ ___ ___ ___ ___

NAME____________________________

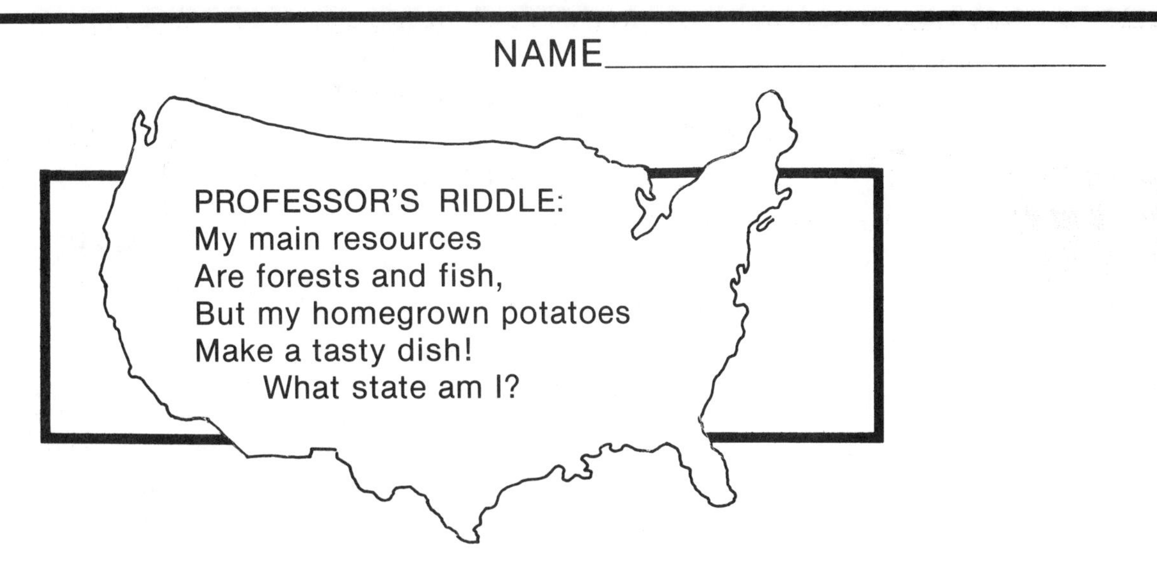

To find out, add. Write the matching letter in the blank above the sum. The letters spell the answer. Some letters will not be used.

ANSWER: ____ ____ ____ ____ ____

1,393 2,076 1,944 1,892 2,396

NAME ______________________

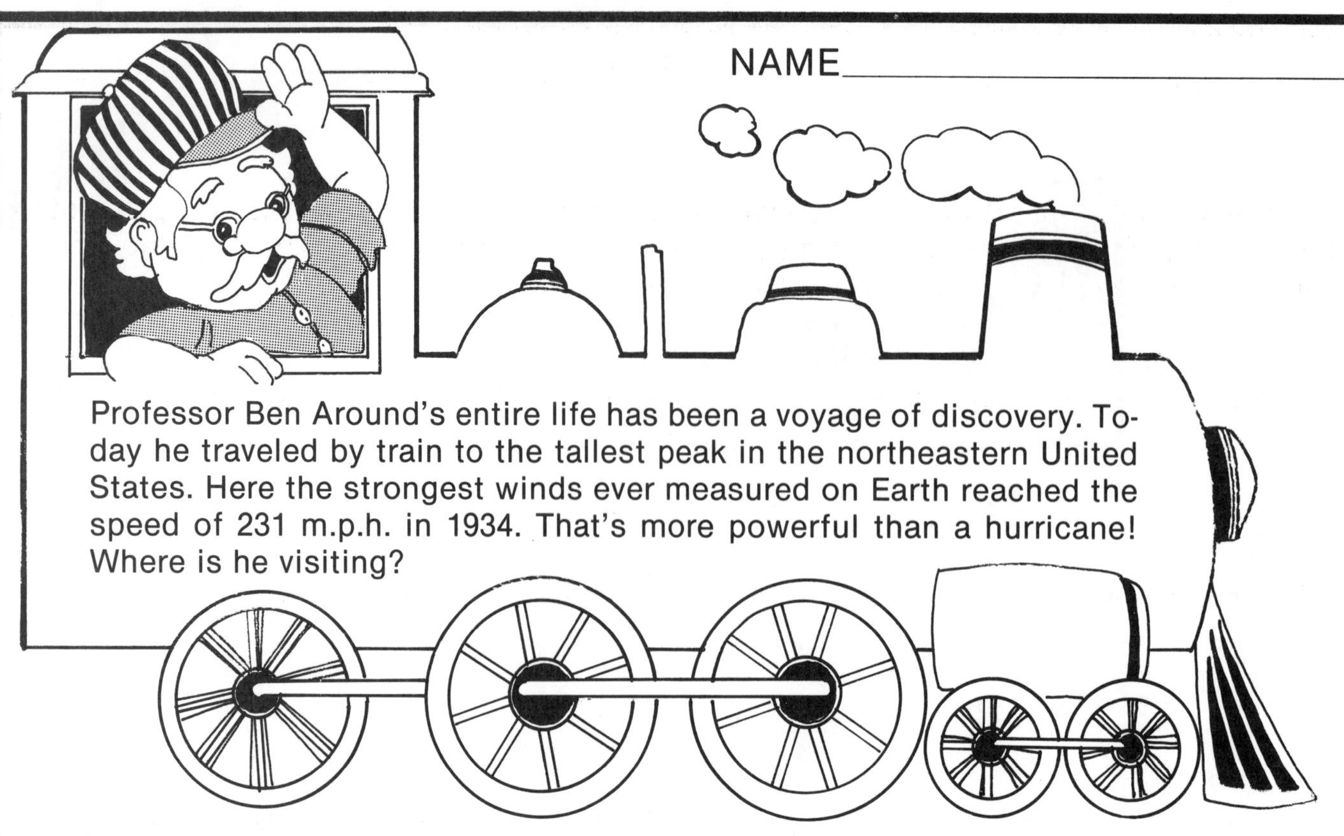

Professor Ben Around's entire life has been a voyage of discovery. Today he traveled by train to the tallest peak in the northeastern United States. Here the strongest winds ever measured on Earth reached the speed of 231 m.p.h. in 1934. That's more powerful than a hurricane! Where is he visiting?

To find out, add. Write the matching letter in the blank above the number of the problem. The letters spell the answer.

1.	2.	3.	4.	5.	6.	7.	8.
5,892	3,576	4,014	2,376	3,045	8,961	5,672	1,593
4,283	8,423	3,742	4,298	6,111	3,514	3,326	6,143
+ 826	+ 3,628	+ 7,516	+ 1,521	+ 1,421	+ 1,562	+ 1,567	+ 2,316

9.	10.	11.	12.	13.	14.
5,278	5,957	6,581	2,866	4,780	1,059
1,507	8,584	3,456	5,623	3,811	6,193
+ 2,237	+ 1,086	+ 4,782	+ 1,563	+ 1,461	+ 6,785

ANSWER:

__ __ . __ __ __ __ __ __ __ __ __ __ , __ __

1 2 . 3 4 5 6 7 8 9 10 11 12 , 13 14

DECODER

A	O	I	T	N	S	M	G	H	W
8,195	14,819	10,565	15,627	10,052	10,577	11,001	9,022	14,037	15,272

NAME______________________________

Professor Ben Around was hit by a wave of excitement as he stood in the only place in the United States where four states meet. The states are New Mexico, Arizona, Utah, and Colorado. What is the name of this place in New Mexico?

To find out, add. Write the matching letter in the blank above the sum. Some letters are used more than once. The letters spell the answer.

N
3,452 + 6,531 = ________

E
76,822 + 59,291 = ________

T
6,231 + 12,123 = ________

U
2,687 + 4,560 = ________

R
6,140 + 2,365 = ________

O
4,625 + 2,586 = ________

C
36,287 + 13,508 = ________

S
15,726 + 8,921 = ________

M
28,815 + 4,876 = ________

F
87,249 + 61,187 = ________

ANSWER:

___	___	___	___		___	___	___	___	___	___	___
148,436	7,211	7,247	8,505		49,795	7,211	8,505	9,983	136,113	8,505	24,647

___	___	___	___	___	___	___	___
33,691	7,211	9,983	7,247	33,691	136,113	9,983	18,354

NAME________________________

To find out, add. Cross out the letter above each sum in the decoder. Unscramble the remaining letters to spell the answer. Write the letters in the blanks below.

9,628	7,836	9,076	6,275
317	3,541	1,367	2,836
8,329	2,062	8,625	3,096
+ 5,641	+ 1,934	+ 4,330	+ 6,055

8,327	8,096	5,970	3,612
8,099	7,412	2,785	3,495
8,205	5,112	7,256	7,221
+ 4,987	+ 2,253	+ 2,316	+ 3,148

DECODER

U 18,262	R 14,218	N 19,785	E 78,321	V 13,721
D 23,915	S 18,327	I 17,476	O 56,542	W 22,873
K 15,373	M 63,498	T 98,971	A 29,618	H 23,398

ANSWER: ___ ___ ___ ___ ___ ___ ___

NAME________________________

Hello friends,

Navigating down the Charles River is a wonderful way to spend the afternoon. Today my nephew, Ben There, and I enjoyed racing our boat against a Harvard team. We are in the famous city which hosts Harvard University. What city are we visiting?

Professor Ben Around

To find out, add. Write the matching letter beside each sum on the sail. The letters spell the answer. Write the letters in the blanks below.

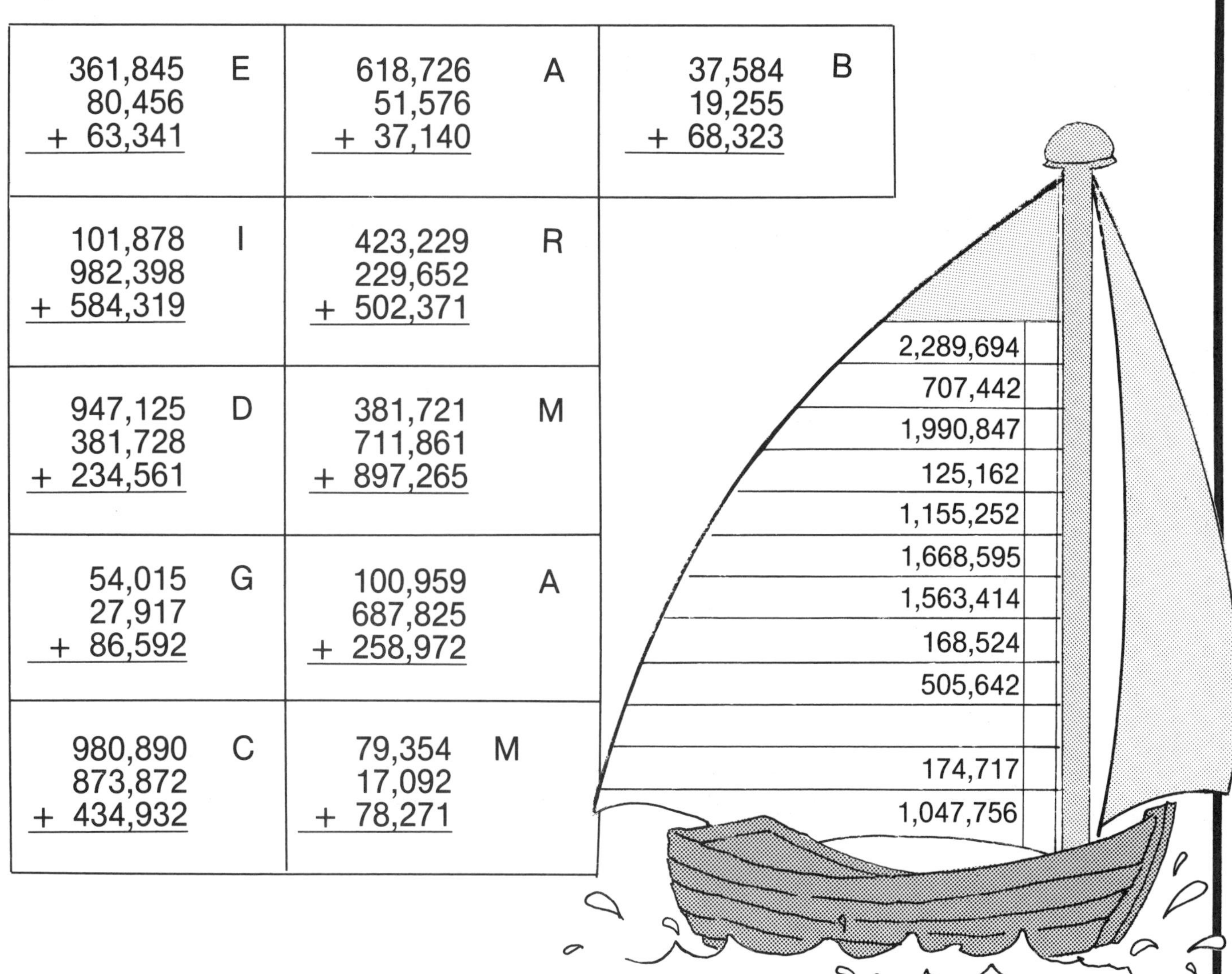

361,845 80,456 + 63,341 E	618,726 51,576 + 37,140 A	37,584 19,255 + 68,323 B
101,878 982,398 + 584,319 I	423,229 229,652 + 502,371 R	
947,125 381,728 + 234,561 D	381,721 711,861 + 897,265 M	
54,015 27,917 + 86,592 G	100,959 687,825 + 258,972 A	
980,890 873,872 + 434,932 C	79,354 17,092 + 78,271 M	

ANSWER: _ _ _ _ _ _ _ _ _ _ _ _ _ _, _ _ _ _

NAME________________________

To find out, add. Shade in each sum below that is divisible by three. Write the matching letters in the blanks above the sums. Some letters are used more than once. The letters spell the answer.

HINT: A number is divisible by three if the sum of the numbers can be divided by three.

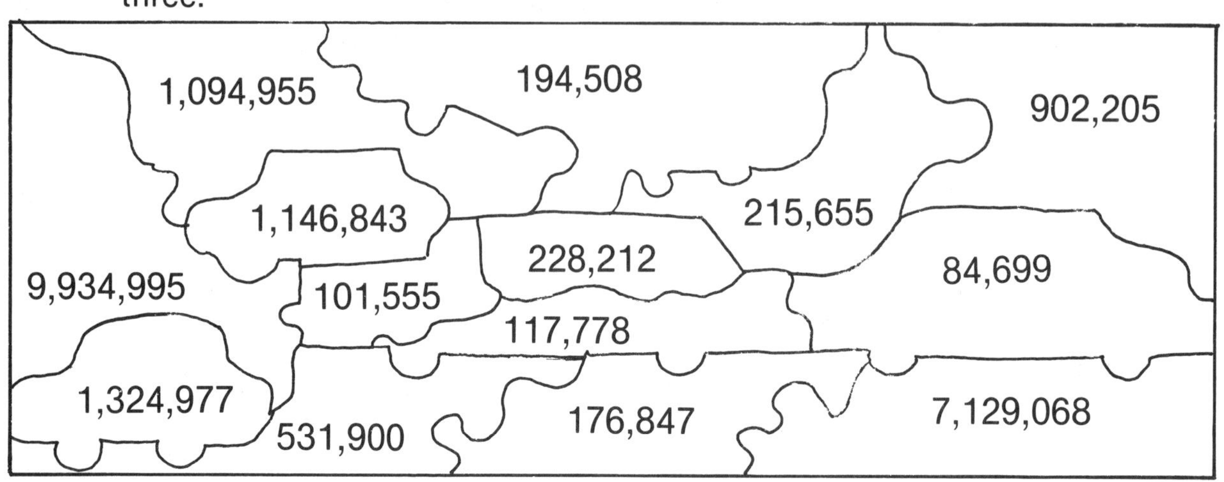

ANSWER:

___ ___ ___ ___ ___ ___ ___ ___ ___ ___
9,934,995 7,129,068 1,146,843 1,146,843 531,900 215,655 194,508 176,847 176,847 531,900

___ ___ ___ ___   ___ ___ ___ ___ ___
1,094,955 1,324,977 176,847 902,205   84,699 176,847 1,324,977 902,205 1,094,955

E	136,754 + 369,611 + 25,535 =	**N**	656,778 + 326,542 + 163,523 =
R	100,863 + 55,301 + 72,048 =	**O**	625,270 + 2,341,625 + 4,162,173 =
S	243,279 + 569,652 + 282,024 =	**P**	62,980 + 29,029 + 25,769 =
A	352,576 + 247,200 + 725,201 =	**B**	4,251,432 + 4,341,251 + 1,342,312 =
L	12,485 + 74,593 + 89,769 =	**G**	18,042 + 40,682 + 42,831 =
T	785,621 + 56,869 + 59,715 =	**V**	49,568 + 82,491 + 83,596 =
I	35,978 + 88,749 + 69,781 =	**F**	9,562 + 29,513 + 45,624 =

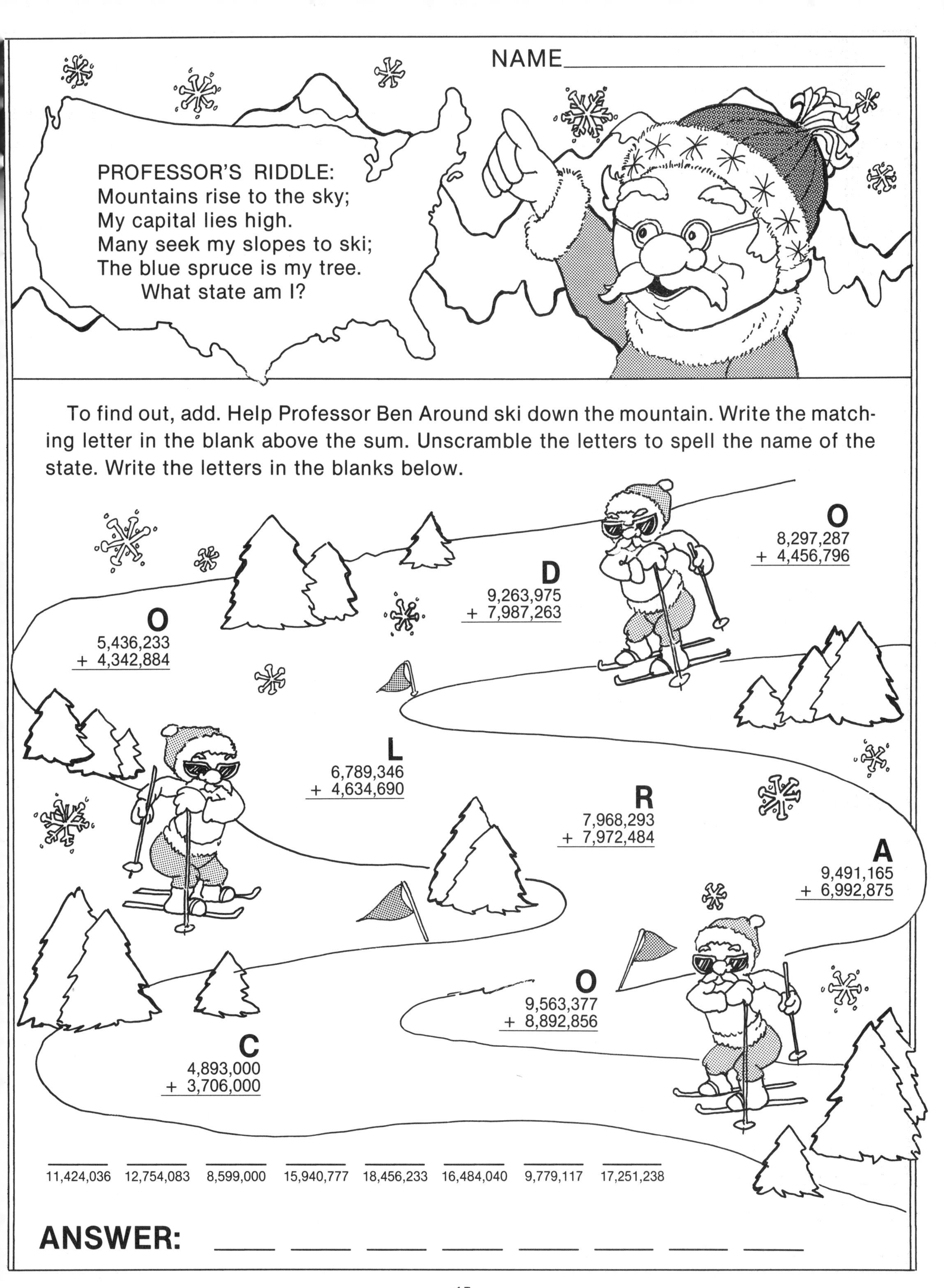
NAME
PROFESSOR'S RIDDLE:
Mountains rise to the sky;
My capital lies high.
Many seek my slopes to ski;
The blue spruce is my tree.
What state am I?
To find out, add. Help Professor Ben Around ski down the mountain. Write the matching letter in the blank above the sum. Unscramble the letters to spell the name of the state. Write the letters in the blanks below.
O
8,297,287
+ 4,456,796
D
9,263,975
+ 7,987,263
O
5,436,233
+ 4,342,884
L
6,789,346
+ 4,634,690
R
7,968,293
+ 7,972,484
A
9,491,165
+ 6,992,875
O
9,563,377
+ 8,892,856
C
4,893,000
+ 3,706,000
11,424,036 12,754,083 8,599,000 15,940,777 18,456,233 16,484,040 9,779,117 17,251,238
ANSWER:

NAME ______________________

PROFESSOR'S MAP MYSTERY:
I'm the Maryland track that horses grace
When they run the Preakness race.
What is my name?

To find out, subtract. Cross out the letter below the matching difference in the decoder. The remaining letters spell the answer. Write the letters in the blanks below.

DECODER

512 R	409 O	340 P	605 Y	725 A	692 I	587 L	648 P	438 M	511 R	316 I
558 X	427 T	516 R	680 L	520 A	351 I	460 C	302 C	669 K	598 O	287 S

1. $543 - 27$
2. $630 - 43$
3. $486 - 59$
4. $311 - 24$
5. $543 - 83$
6. $594 - 74$
7. $734 - 86$
8. $603 - 91$
9. $546 - 35$
10. $745 - 76$
11. $653 - 48$
12. $465 - 56$
13. $792 - 67$
14. $630 - 72$
15. $359 - 43$

ANSWER: ___ ___ ___ ___ ___ ___ ___

NAME____________________________

PROFESSOR'S RIDDLE:
Stately beauties
Adorn this place.
The nation watches
Each pretty face.
What is this city?

To find out, subtract. Circle the letter beside the difference in the decoder. The circled letters spell the answer. Write the letters in the blanks below.

DECODER

A	5,407
L	3,815
T	6,687
D	4,051
E	6,238
L	4,829
P	5,156
A	6,559
N	7,138
B	8,250
A	5,387
M	9,123
T	7,475
A	8,810
I	7,789
C	6,759
S	6,502
C	8,829
O	7,394
L	5,285
I	6,416
N	4,300
T	5,649
Y	7,387
S	6,375
N	4,556
M	9,207
J	5,440

1. $8{,}243 - 768$
2. $7{,}091 - 532$
3. $9{,}316 - 487$
4. $6{,}248 - 808$
5. $7{,}352 - 936$
6. $8{,}135 - 748$
7. $5{,}021 - 465$
8. $8{,}316 - 527$
9. $6{,}214 - 807$
10. $7{,}321 - 634$
11. $5{,}432 - 603$
12. $8{,}053 - 915$
13. $6{,}178 - 529$
14. $7{,}654 - 895$

ANSWER: _ _ _ _ _ _ _ _ _ _   _ _ _ _ _ _,   _ _

NAME ______________________

Professor Ben Around has butterflies in his stomach! He and his nephew, Ben There, are in the "Hit Recording Capital of the World" to record their first album. Do you know where they are?

To find out, subtract. Circle the letter beside the difference in the decoder. The circled letters spell the answer. Write the letters in the blanks below.

1. $9{,}243 - 2{,}764$

2. $8{,}640 - 4{,}312$

3. $5{,}304 - 4{,}136$

4. $7{,}381 - 5{,}456$

5. $4{,}351 - 1{,}789$

6. $8{,}743 - 5{,}286$

7. $9{,}051 - 3{,}682$

8. $4{,}507 - 2{,}384$

9. $6{,}321 - 5{,}078$

10. $5{,}942 - 2{,}768$

11. $8{,}211 - 3{,}428$

12. $3{,}215 - 1{,}758$

13. $5{,}830 - 3{,}467$

14. $6{,}502 - 4{,}376$

DECODER

T	3,815
M	4,328
H	2,873
U	1,243
S	2,123
I	4,057
C	5,369
N	3,024
R	4,879
L	1,168
E	4,783
S	2,363
K	5,439
H	3,457
W	2,870
O	1,925
A	6,479
Y	5,387
L	2,126
S	3,174
A	1,457
V	4,096
L	2,562
B	3,683

ANSWER:

_ _ _ _ _ _ _   _ _ _ _ _ _ _ _ _ , _ _

NAME________________________________

Professor Ben Around is visiting the only major diamond field in the United States. Someone once found a diamond there worth a quarter of a million dollars! Maybe Ben will get lucky, too. Where is Professor Ben?

To find out, subtract. Write the matching letter in the blank above the difference. The letters spell the answer.

ANSWER:

___ ___ ___ ___ ___ ___   ___ ___
86 219 359 249 179 313   274 254

___ ___ ___ ___ ___ ___ ___ ___,
335 191 226 256 244 165 367 326

___ ___ ___ ___ ___ ___ ___ ___
557 73 436 315 322 321 284 188

1. 596 − 342 =	F ______	13. 726 − 405 =	S ______
2. 718 − 403 =	A ______	14. 492 − 125 =	D ______
3. 602 − 516 =	C ______	15. 508 − 234 =	O ______
4. 754 − 318 =	K ______	16. 362 − 183 =	E ______
5. 428 − 106 =	N ______	17. 721 − 408 =	R ______
6. 524 − 275 =	T ______	18. 527 − 362 =	N ______
7. 815 − 258 =	A ______	19. 431 − 205 =	A ______
8. 392 − 173 =	R ______	20. 765 − 521 =	O ______
9. 464 − 208 =	M ______	21. 463 − 128 =	D ______
10. 693 − 502 =	I ______	22. 560 − 234 =	S ______
11. 708 − 635 =	R ______	23. 812 − 528 =	A ______
12. 523 − 164 =	A ______	24. 391 − 203 =	S ______

NAME________________________

To find out, subtract and round each number to the nearest ten. Write the matching letter in the blank above the number of the problem. The letters spell the answer.

ANSWER:

__ __ __ __ __ __ __ __ __ __ , __ __
1 2 3 4 5 6 7 8 9 10 , 11 12

1. $178 - 32$	2. $346 - 58$	3. $654 - 29$	4. $586 - 134$
5. $893 - 201$	6. $738 - 249$	7. $523 - 147$	8. $486 - 275$
9. $310 - 176$	10. $625 - 358$	11. $703 - 462$	12. $685 - 403$

DECODER

L	H	L	A	E	L	U	I	T	V	N	S
130	150	210	240	270	280	290	380	450	490	630	690

NAME____________________

Hello Boys and Girls,

I'm visiting the only city in the United States where I can look directly south into Canada. The city is the oldest major city in the Midwest. Can you identify it?

Professor Ben Around

To find out, subtract. Draw a line from the letter beside each problem to the difference. Write the letter in the blank beside the difference. The letters spell the answer. Write the letters in the blanks below.

Problem	Letter	Difference	
1. 6,871 − 2,345 =	A	2,903	______
2. 2,564 − 902 =	R	931	______
3. 4,872 − 2,468 =	G	861	______
4. 1,543 − 682 =	T	1,662	______
5. 4,106 − 829 =	I	1,912	______
6. 4,017 − 2,591 =	I	3,277	______
7. 6,083 − 2,519 =	H	3,304	______
8. 5,164 − 3,406 =	I	3,083	______
9. 3,942 − 638 =	T	1,426	______
10. 1,760 − 829 =	E	3,959	______
11. 2,346 − 434 =	O	3,564	______
12. 5,235 − 1,276 =	C	1,758	______
13. 3,986 − 1,392 =	N	2,404	______
14. 3,697 − 794 =	D	4,526	______
15. 7,291 − 4,208 =	M	2,594	______

ANSWER: _ _ _ _ _ _ _, _ _ _ _ _ _ _ _

NAME________________________

Uncle Go Go and Professor Ben Around love camping under the stars. Their tent is pitched in eastern California near El Capitan, a granite mass rising 3,600 feet. Can you identify their campsite?

To find out, subtract. Find the difference on the star and write the matching letter in the blank above the number of the problem. The letters spell the answer.

ANSWER:

__ __ __ __ __ __ __ __
1 2 3 4 5 6 7 8

__ __ __ __ __ __ __ __   __ __ __ __
9 10 11 12 13 14 15 16   17 18 19 20

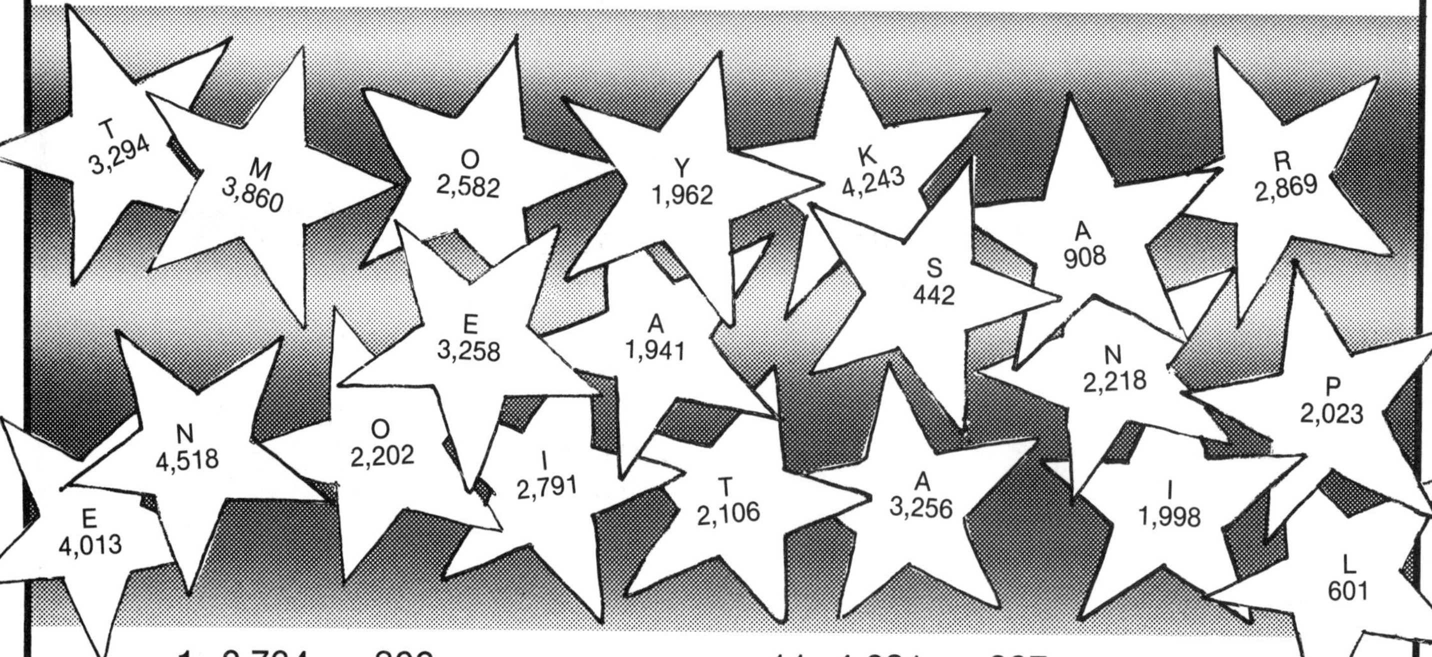

1. 2,764 − 802 = ______
2. 3,518 − 936 = ______
3. 1,023 − 581 = ______
4. 4,961 − 948 = ______
5. 4,653 − 793 = ______
6. 2,867 − 869 = ______
7. 3,057 − 951 = ______
8. 4,023 − 765 = ______
9. 2,961 − 743 = ______
10. 1,543 − 635 = ______
11. 4,281 − 987 = ______
12. 3,652 − 861 = ______
13. 2,860 − 658 = ______
14. 5,430 − 912 = ______
15. 2,348 − 407 = ______
16. 1,259 − 658 = ______
17. 2,765 − 742 = ______
18. 4,081 − 825 = ______
19. 3,807 − 938 = ______
20. 5,086 − 843 = ______

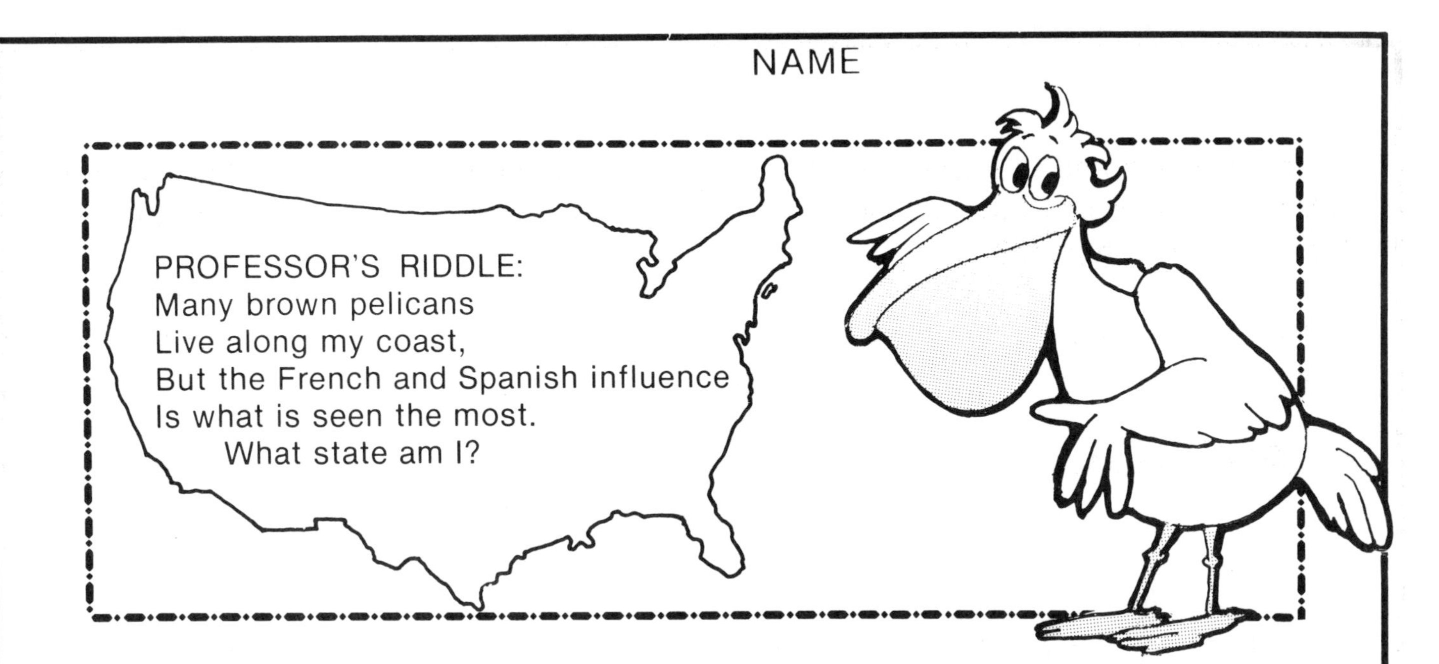

To find out, subtract. Cross out the letter beside each difference in the decoder. The remaining letters spell the answer. Write the letters in the blanks below.

1. $\begin{array}{r} 9{,}241 \\ -\ 6{,}852 \\ \hline \end{array}$

2. $\begin{array}{r} 7{,}615 \\ -\ 4{,}724 \\ \hline \end{array}$

3. $\begin{array}{r} 8{,}123 \\ -\ 5{,}345 \\ \hline \end{array}$

4. $\begin{array}{r} 6{,}432 \\ -\ 3{,}617 \\ \hline \end{array}$

5. $\begin{array}{r} 4{,}219 \\ -\ 1{,}506 \\ \hline \end{array}$

6. $\begin{array}{r} 9{,}016 \\ -\ 2{,}438 \\ \hline \end{array}$

7. $\begin{array}{r} 7{,}438 \\ -\ 4{,}176 \\ \hline \end{array}$

8. $\begin{array}{r} 5{,}213 \\ -\ 3{,}623 \\ \hline \end{array}$

9. $\begin{array}{r} 7{,}290 \\ -\ 3{,}485 \\ \hline \end{array}$

10. $\begin{array}{r} 8{,}910 \\ -\ 5{,}629 \\ \hline \end{array}$

11. $\begin{array}{r} 6{,}093 \\ -\ 4{,}178 \\ \hline \end{array}$

12. $\begin{array}{r} 8{,}073 \\ -\ 4{,}254 \\ \hline \end{array}$

DECODER

R	2,713
S	3,819
L	4,605
N	2,891
O	3,410
G	1,915
W	2,815
U	3,297
T	3,281
B	2,389
I	7,604
S	5,823
H	3,262
M	1,590
I	3,173
D	2,778
A	4,825
Y	6,578
N	5,923
C	3,805
A	4,074

ANSWER: __ __ __ __ __ __ __ __ __

NAME________________________

PROFESSOR'S MAP MYSTERY:
I'm America's tallest lighthouse
Warning the ships at sea
In the graveyard of the Atlantic.
Beware! Stay away from me!
What is my location?

To find out, subtract. Circle the letter above the difference. The circled letters spell the answer. Write the letters in the blanks below.

1.	6,432 − 804 =	2.	5,290 − 765 =	3.	8,012 − 934 =
	C N		E A		W P
	5,628 5,622		4,435 4,525		7,086 7,078
4.	7,691 − 543 =	5.	6,342 − 721 =	6.	9,307 − 834 =
	E Y		O H		R A
	7,148 7,142		5,661 5,621		8,373 8,473
7.	7,620 − 541 =	8.	8,030 − 850 =	9.	7,196 − 430 =
	T K		H T		E B
	7,079 6,979		7,080 7,180		6,766 6,786
10.	4,261 − 538 =	11.	6,573 − 842 =	12.	8,399 − 763 =
	A R		A R		B S
	3,733 3,723		5,731 5,621		7,546 7,636
13.	6,035 − 786 =	14.	5,821 − 907 =		
	N O		R C		
	5,249 5,139		4,921 4,914		

ANSWER: _ _ _ _ _ _ _ _ _ _ _ _ _ _ _, _ _

NAME ____________________

TOUCHDOWN! While visiting the oldest city in the Badger State, Professor Ben Around watched his favorite football team score an exciting victory. Do you know what city the professor was visiting?

To find out, subtract. Circle the letter beside the difference in the decoders. The circled letters spell the answer. Write the letters in the blanks below.

DECODER

G	2,176
A	1,135
R	1,008
T	805
E	608
L	938
E	777
N	1,405
U	2,534
B	1,922
R	3,027
A	1,847
S	1,950
Y	1,096

1. 596 less than 1,204 = ______
2. 382 less than 858 = ______
3. 691 less than 2,538 = ______
4. 764 less than 1,926 = ______
5. 808 less than 3,105 = ______
6. 497 less than 2,673 = ______
7. 548 less than 3,469 = ______
8. 348 less than 2,804 = ______
9. 672 less than 1,680 = ______
10. 836 less than 3,458 = ______
11. 745 less than 1,943 = ______
12. 521 less than 2,459 = ______
13. 298 less than 1,703 = ______
14. 452 less than 2,374 = ______
15. 581 less than 1,358 = ______
16. 632 less than 2,461 = ______
17. 490 less than 1,586 = ______

DECODER

A	852
W	1,162
R	907
I	476
K	1,652
S	2,921
C	2,456
E	3,091
O	1,198
N	1,829
T	764
S	1,938
I	2,297
N	2,622

ANSWER: _ _ _ _ _ _ _ _ _, _ _ _ _ _ _ _ _

To find out, subtract. Shade the spaces below that contain the answers to the problems. The answer to the riddle appears in the drawing.

1. $\begin{array}{r} 26,401 \\ -\ 18,328 \\ \hline \end{array}$ 2. $\begin{array}{r} 49,764 \\ -\ 32,187 \\ \hline \end{array}$ 3. $\begin{array}{r} 34,801 \\ -\ 19,243 \\ \hline \end{array}$ 4. $\begin{array}{r} 32,465 \\ -\ 17,374 \\ \hline \end{array}$ 5. $\begin{array}{r} 46,286 \\ -\ 34,815 \\ \hline \end{array}$

6. $\begin{array}{r} 64,807 \\ -\ 17,928 \\ \hline \end{array}$ 7. $\begin{array}{r} 54,353 \\ -\ 32,687 \\ \hline \end{array}$ 8. $\begin{array}{r} 27,384 \\ -\ 19,478 \\ \hline \end{array}$ 9. $\begin{array}{r} 38,209 \\ -\ 21,786 \\ \hline \end{array}$ 10. $\begin{array}{r} 76,093 \\ -\ 51,476 \\ \hline \end{array}$

11. $\begin{array}{r} 64,384 \\ -\ 32,476 \\ \hline \end{array}$ 12. $\begin{array}{r} 28,173 \\ -\ 17,248 \\ \hline \end{array}$ 13. $\begin{array}{r} 24,307 \\ -\ 12,913 \\ \hline \end{array}$ 14. $\begin{array}{r} 31,652 \\ -\ 19,487 \\ \hline \end{array}$

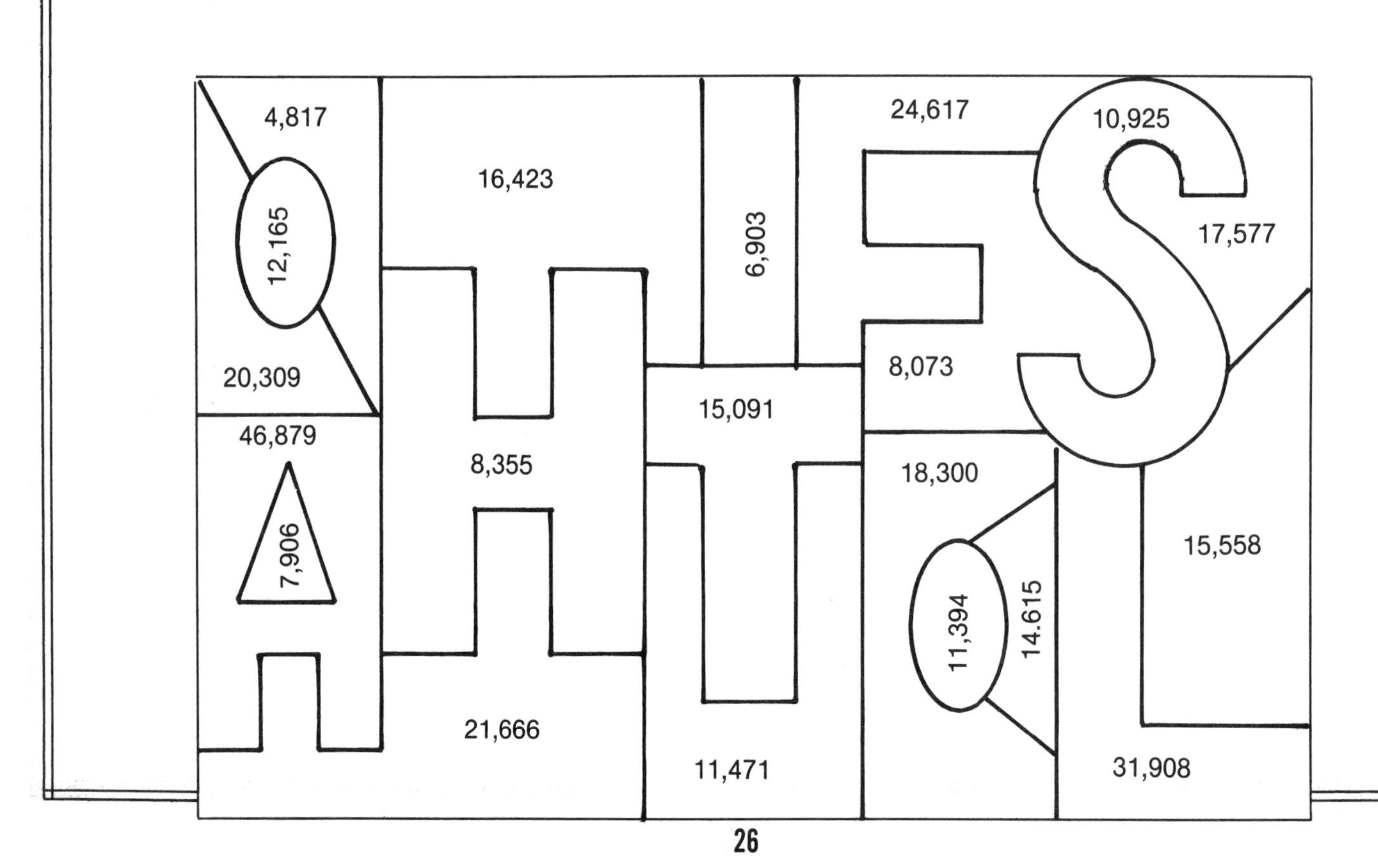

NAME ____________________________

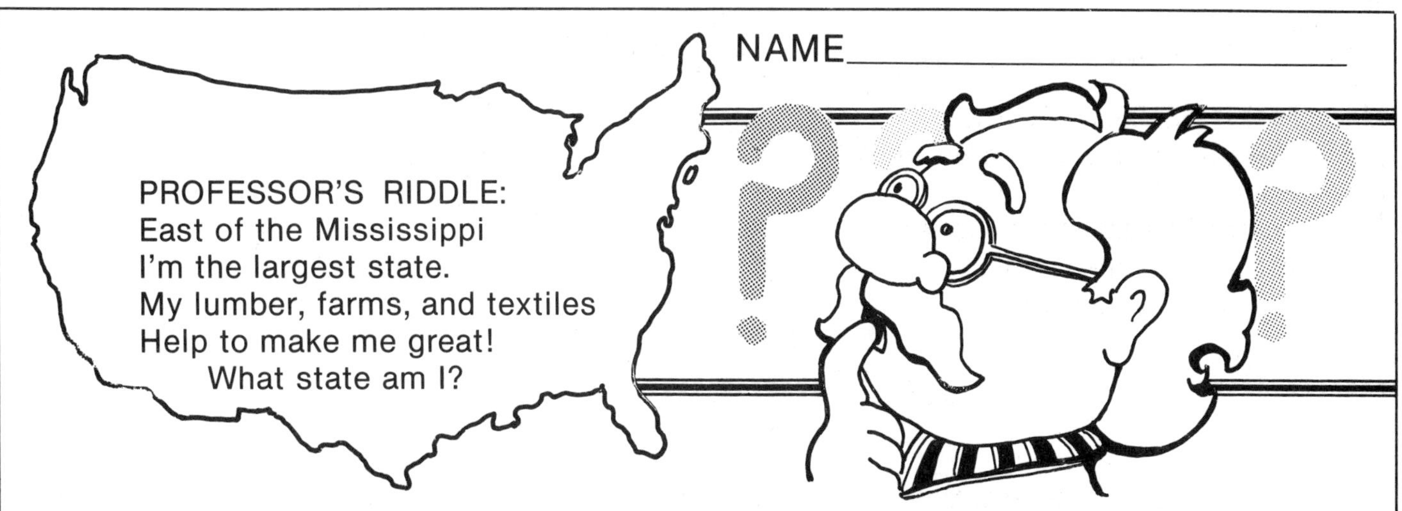

To find out, subtract. Cross out the letter beside each difference that is an odd number. The remaining letters spell the answer. Write the letters in the blanks below.

HINT: An odd number is a whole number that is not divisible by 2. Look at the digit in the ones column to determine if a number is odd or even.

29,402 − 8,615 B	36,143 − 9,204 A	21,046 − 8,212 G	17,432 − 2,175 W
18,476 − 2,514 E	42,817 − 9,456 L	52,178 − 8,415 K	30,172 − 8,043 N
46,253 − 8,198 S	20,185 − 4,561 O	28,436 − 5,102 R	21,730 − 8,147 P
28,134 − 9,020 G	18,760 − 2,918 I	30,175 − 8,176 H	52,186 8,042 A

ANSWER: ___ ___ ___ ___ ___ ___ ___

NAME______________________________

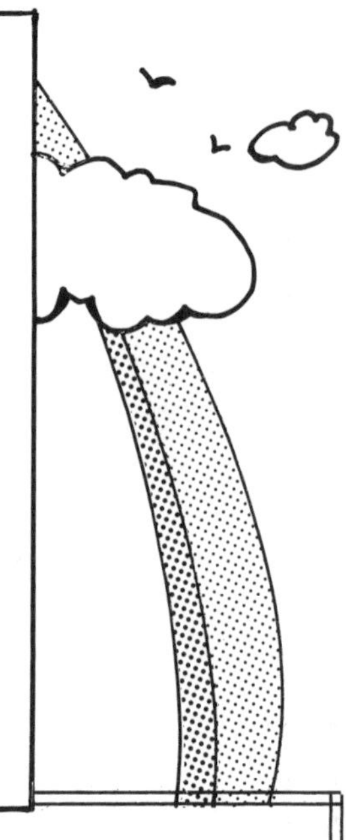

Dear Boys and Girls,

I'm visiting the home of the tallest monument in the United States. When I ride to the top, I can see Uncle Go Go's riverboat sailing down the Mississippi River! What city am I visiting?

Professor Ben Around

To find out, subtract. Write the matching letter in the blank above the difference. Some letters will not be used. The letters spell the answer.

ANSWER: ____ ____ . ____ ____ ____ ____ ____

20,759 17,718 . 16,553 20,561 8,246 17,858 19,588

1. 17,604 − 9,358 U	2. 28,351 − 8,763 S	3. 36,058 − 7,613 B	4. 24,076 − 6,358 T
5. 29,035 − 18,216 W	6. 25,178 − 19,245 R	7. 27,381 − 17,945 Y	8. 52,436 − 31,875 O
9. 46,201 − 28,343 I	10. 28,906 − 12,353 L	11. 34,962 − 28,358 J	12. 44,307 − 21,486 K
13. 27,643 − 18,485 P	14. 40,516 − 26,384 D	15. 31,086 − 25,243 E	16. 36,045 − 15,286 S

PROFESSOR'S MAP MYSTERY:
From my vantage point
You will see
Chattanooga, Tennessee.
What physical feature am I?

NAME________________________

To find out, subtract. Circle the letter below each difference in the decoder. The circled letters from left to right spell the answer. Write the letters in the blanks below.

DECODER

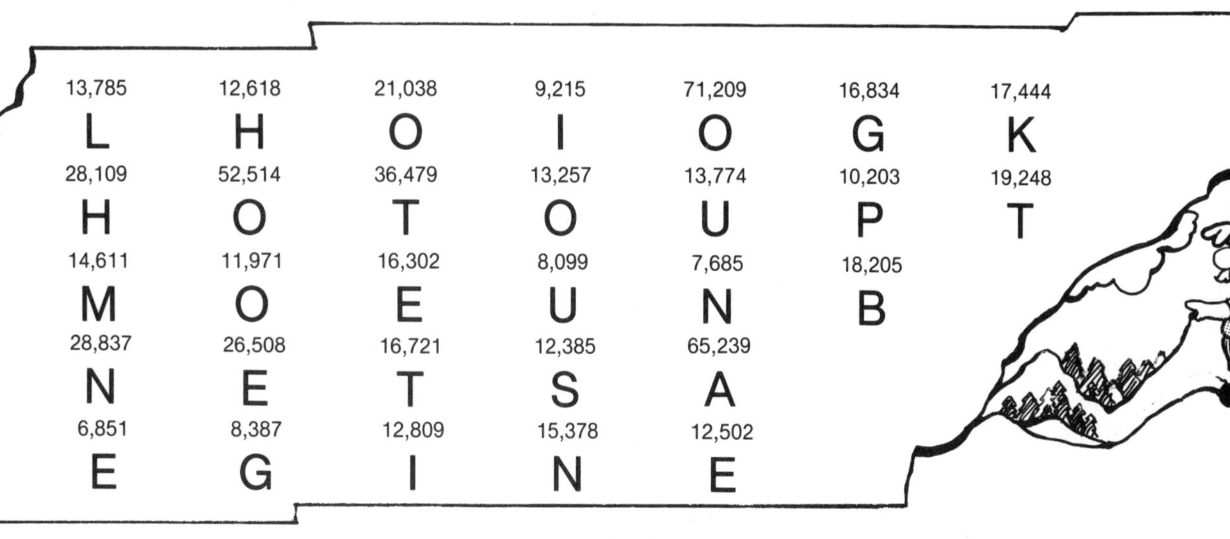

1. 17,304 − 9,205	2. 86,452 − 15,243	3. 26,354 − 10,976
4. 38,072 − 24,287	5. 34,512 − 13,474	6. 90,346 − 25,107
7. 25,241 − 12,432	8. 35,762 − 16,514	9. 23,017 − 8,406
10. 38,754 − 26,783	11. 52,017 − 38,243	12. 87,492 − 34,978
13. 73,104 − 44,267	14. 21,079 − 4,358	15. 43,502 − 26,058

ANSWER: ___ ___ ___ ___ ___ ___ ___

___ ___ ___ ___ ___ ___ ___ ___

NAME________________________

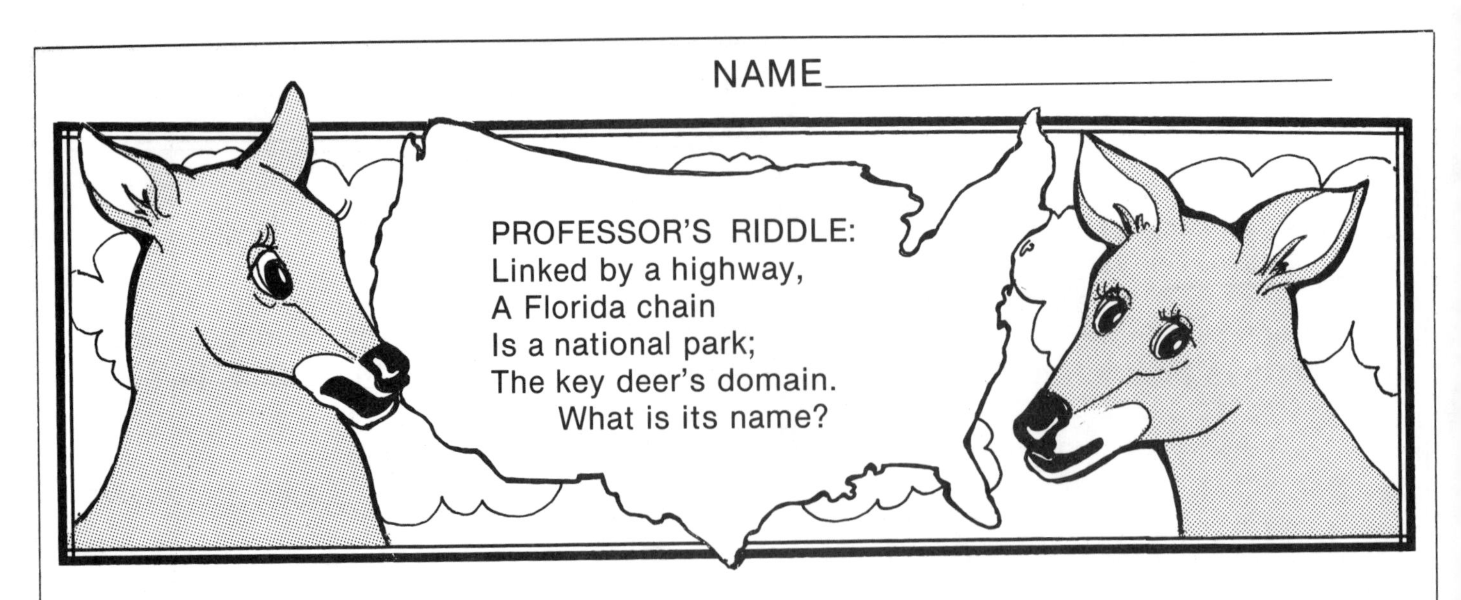

To find out, subtract. Write the matching letter in the blank above the difference. Some letters are used more than once. The letters spell the answer.

ANSWER:

___	___	___	___	___	___	___	___	___	___
371,531	216,703	371,531	424,033	842,848	285,193	558,944	222,152	371,531	216,936

___	___	___	___	___	___	___	___		___	___	___	___
181,908	558,944	245,789	560,908	395,622	181,908	558,944	285,193		439,924	558,944	424,033	167,377

G	871,204 − 28,356	O	432,107 − 36,485	V	293,250 − 76,547
I	654,186 − 93,278	A	576,294 − 17,350	P	472,381 − 32,457
T	543,258 − 297,469	E	389,175 − 17,644	R	876,209 − 452,176
N	652,173 − 470,265	L	520,365 − 235,172	S	703,129 − 486,193
K	912,384 − 745,007	D	376,054 − 153,902		

NAME________________________________

Dear Boys and Girls,

The world is full of remarkable things to see and do. My nephew, Ben There, and some of his friends invited me to join them rafting. We set adrift the largest lake in New Hampshire. Do you know where we were rafting?

Professor Ben Around

To find out, multiply. Write the matching letter in the blank above the product. The letters spell the answer. Some letters will not be used.

ANSWER:

__ __ __ __   __ __ __ __ __ __ __ __ __ __ __ __ __ ,   __ __

189 69 476 166   375 343 72 738 198 171 96 140 208 82 135 132 108   260 85

NAME________________________

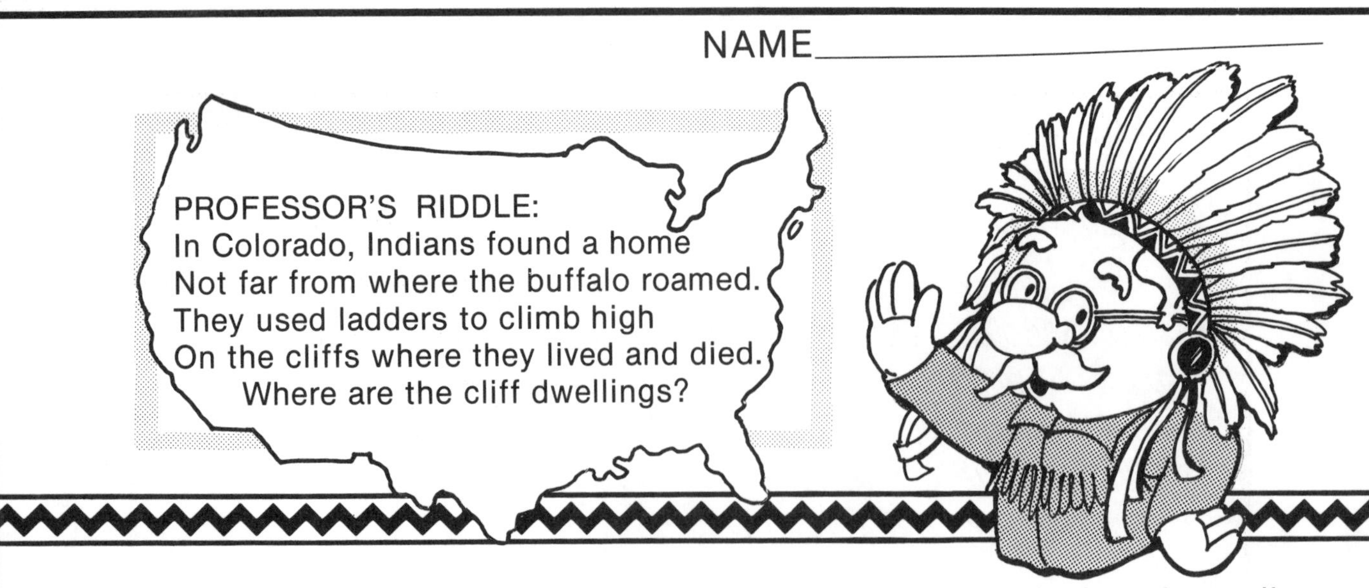

To find out, begin at START and follow the path of multiples of 4 in order until you reach FINISH. Write the letters from the path in the blanks below. The letters spell the answer.

ANSWER: _ _ _ _ _   _ _ _ _ _ _ _

_ _ _ _ _ _ _ _ _ _ _ _ _   _ _ _ _ _

START

4 **M**	8 **E**	9 **T**	6 **P**	47 **I**	13 **K**	14 **H**	23 **G**	29 **D**
3 **R**	12 **S**	11 **B**	7 **W**	2 **F**	15 **C**	31 **E**	18 **J**	46 **S**
5 **F**	16 **A**	20 **V**	25 **D**	43 **R**	45 **O**	47 **M**	49 **S**	50 **L**
17 **P**	9 **U**	24 **E**	10 **Z**	42 **X**	44 **A**	48 **T**	52 **I**	56 **O**
19 **V**	27 **Q**	28 **R**	32 **D**	36 **E**	40 **N**	68 **L**	64 **A**	60 **N**
21 **F**	30 **A**	33 **N**	34 **T**	35 **K**	37 **B**	72 **P**	71 **G**	73 **E**
22 **A**	23 **C**	26 **I**	41 **M**	39 **O**	38 **S**	76 **A**	80 **R**	84 **K**

FINISH

NAME________________________________

Hello Boys and Girls,

It's a unique experience to visit this city because a narrow river meanders over a fifteen-mile course through the heart of the city. A wonderful way to tour the city is by canoe or gondola! Shops and restaurants await you on the River Walk.

Where am I visiting?

Professor Ben Around

To find out, multiply. Write the matching letter in the blank above the product. The letters spell the answer.

ANSWER:

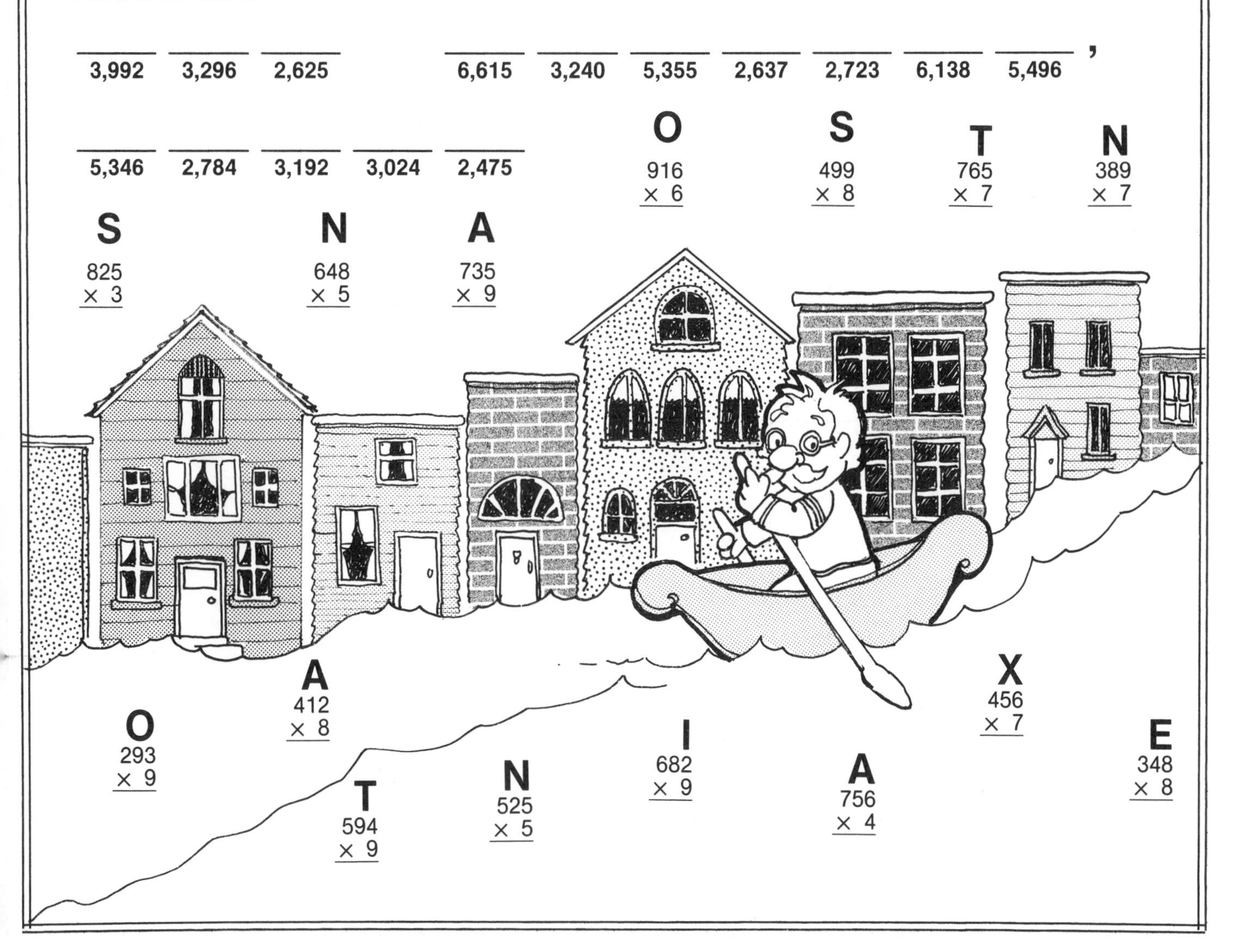

To find out, multiply. Write the matching letter in the blank above the product. The letters spell the answer.

ANSWER: ___ ___ ___ ___ ___ ___ ___ ___ ,

15,345 63,536 9,741 42,714 32,418 15,282 25,024 36,799

___ ___ ___ ___ ___ ___ ___

42,848 34,825 39,316 16,780 28,476 11,416 5,148

Y	M	N	W	G	O	I
4,975 × 7	3,356 × 5	1,427 × 8	5,356 × 8	1,287 × 4	9,829 × 4	3,164 × 9

H	N	Y	E	E	C	N	E
7,942 × 8	2,547 × 6	6,102 × 7	3,247 × 3	5,257 × 7	1,705 × 9	6,256 × 4	3,602 × 9

PROFESSOR'S RIDDLE:
John Kennedy was born here;
The Atlantic Ocean is very near.
This was the home of Pilgrims' pride;
Also the site of Paul Revere's ride!
What is this state?

NAME________________________

To find out, multiply. Find the product in the decoder and write the matching letter in the outside circle beside the problem. The letters spell the answer. Write the letters in the blanks below.

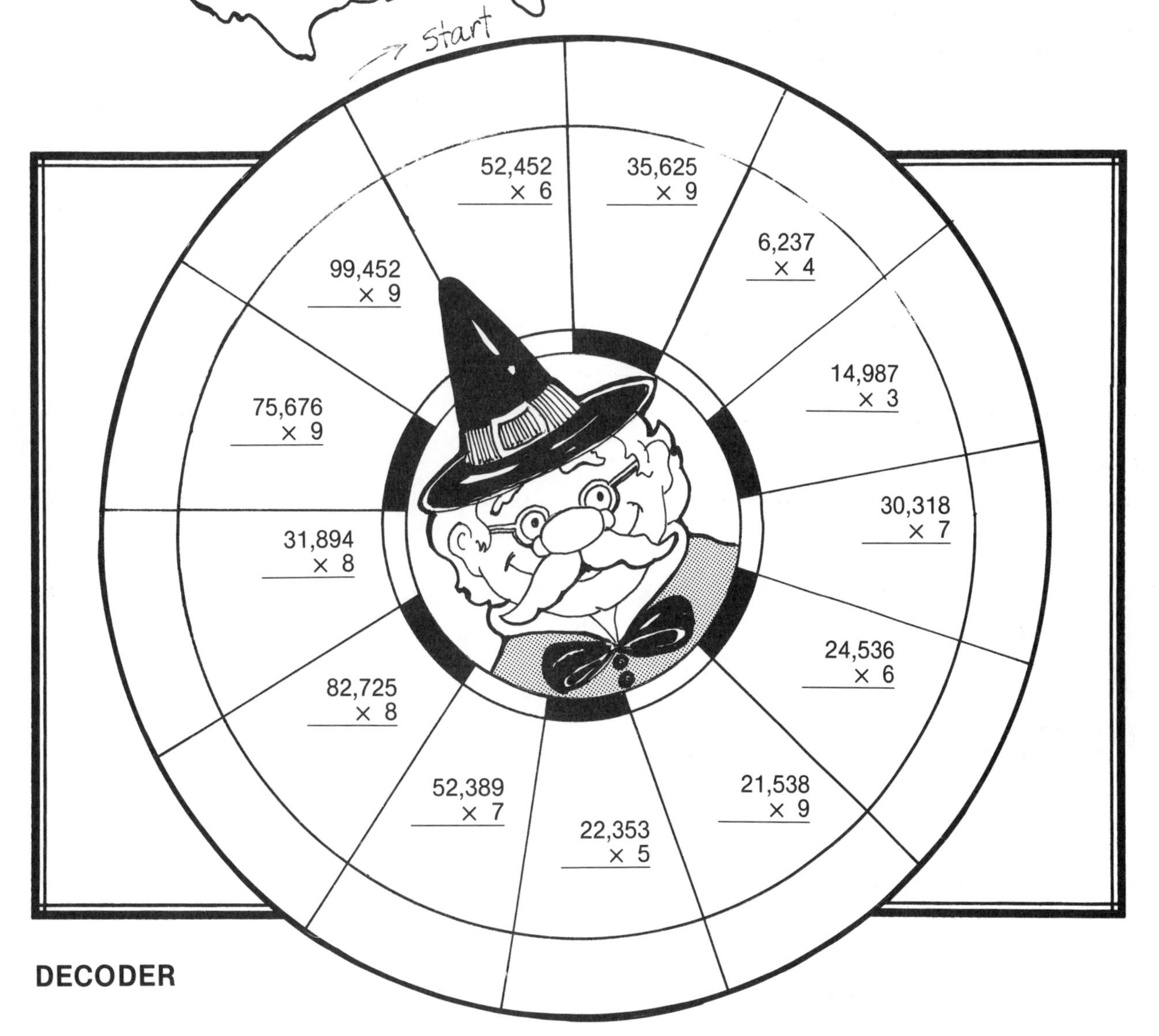

DECODER

S = 895,068	T = 255,152	A = 320,625
U = 111,765	H = 193,842	S = 44,961
T = 681,084	C = 147,216	E = 661,800
S = 366,723	S = 24,948	
A = 212,226	M = 314,712	

ANSWER: __ __ __ __ __ __ __ __ __ __ __ __ __

NAME________________________

PROFESSOR'S MAP MYSTERY:
I'm covered with terraces and hills;
My brilliant hues bring people chills.
With colorful rocks and sand,
Arizona is a beautiful land.
What am I?

To find out, multiply. Write the matching letter in the blank above the number of each problem. The letters spell the answer.

1. 82,835 × 5 = ________
2. 78,523 × 6 = ________
3. 67,589 × 8 = ________
4. 45,324 × 9 = ________
5. 33,013 × 3 = ________
6. 29,240 × 4 = ________
7. 83,328 × 7 = ________
8. 92,119 × 2 = ________
9. 77,832 × 8 = ________
10. 63,385 × 4 = ________
11. 18,475 × 9 = ________
12. 29,177 × 6 = ________
13. 91,586 × 8 = ________

ANSWER:

__ __ __ __ __ __ __   __ __ __ __ __ __
1 2 3 4 5 6 7   8 9 10 11 12 13

DECODER

E	S	T	D	R	P	E	N	A	E	T	I	D
166,275	253,540	732,688	184,238	175,062	414,175	622,656	407,916	471,138	116,960	99,039	540,712	583,296

NAME ____________________

PROFESSOR'S RIDDLE:
Farms cover most of this state,
Growing foods to put on our plates.
Raising corn is a very big deal
In a state famous for Buffalo Bill.
What state is this?

To find out, multiply. Cross out the letter beside the product in the decoder. The remaining letters spell the answer. Write the letters in the blanks below.

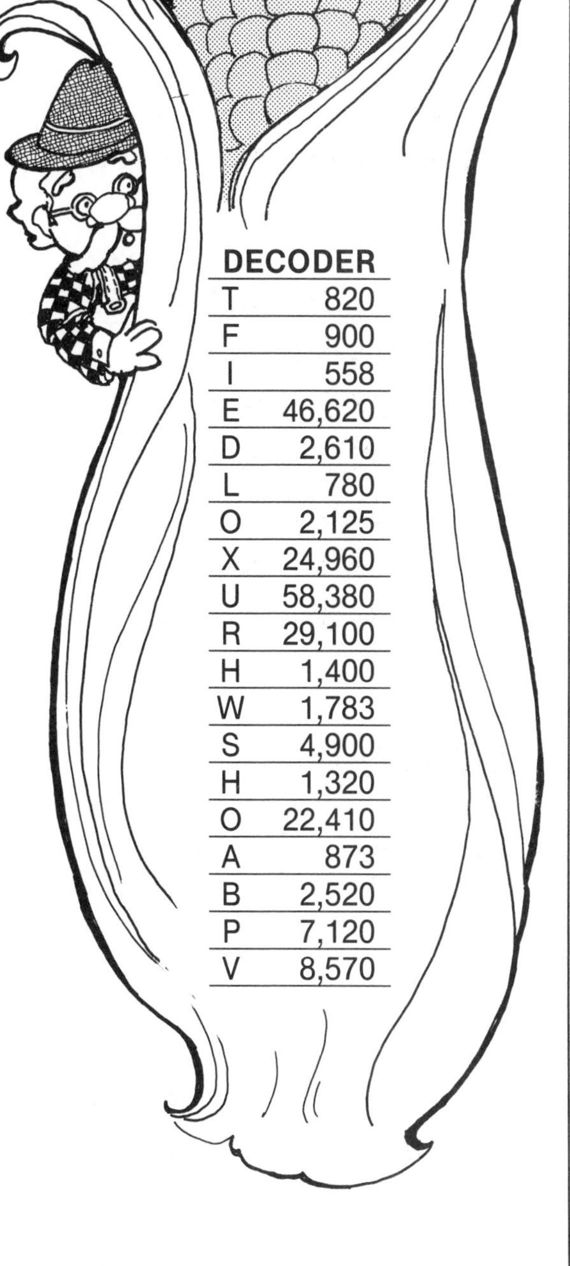

DECODER	
T	820
F	900
I	558
E	46,620
D	2,610
L	780
O	2,125
X	24,960
U	58,380
R	29,100
H	1,400
W	1,783
S	4,900
H	1,320
O	22,410
A	873
B	2,520
P	7,120
V	8,570

582×50 44×30 666×70 35×40 78×10

973×60 89×80 29×90 747×30 18×50

857×10 41×20 63×40 416×60 70×70

ANSWER: ____ ____ ____ ____

NAME________________________

To find out, multiply. Circle the letter above each product in the decoder. Connect the circled letters in order of the problems. The letters along the path of products spell the answer. Write the letters in the blanks below.

1. 75×28

2. 84×54

3. 92×45

4. 68×46

5. 62×29

6. 58×42

7. 82×86

8. 26×29

9. 25×47

10. 39×58

11. 18×67

DECODER

A 899	D 4,140	H 3,827	B 2,905
O 4,536	I 1,127	G 3,128	C 2,436
M 1,463	START D 2,100	E 1,798	I 7,052
FINISH S 1,206	K 2,262	Y 1,175	T 754

ANSWER: __ __ __ __ __ __ __ __ __, __ __

NAME ____________________

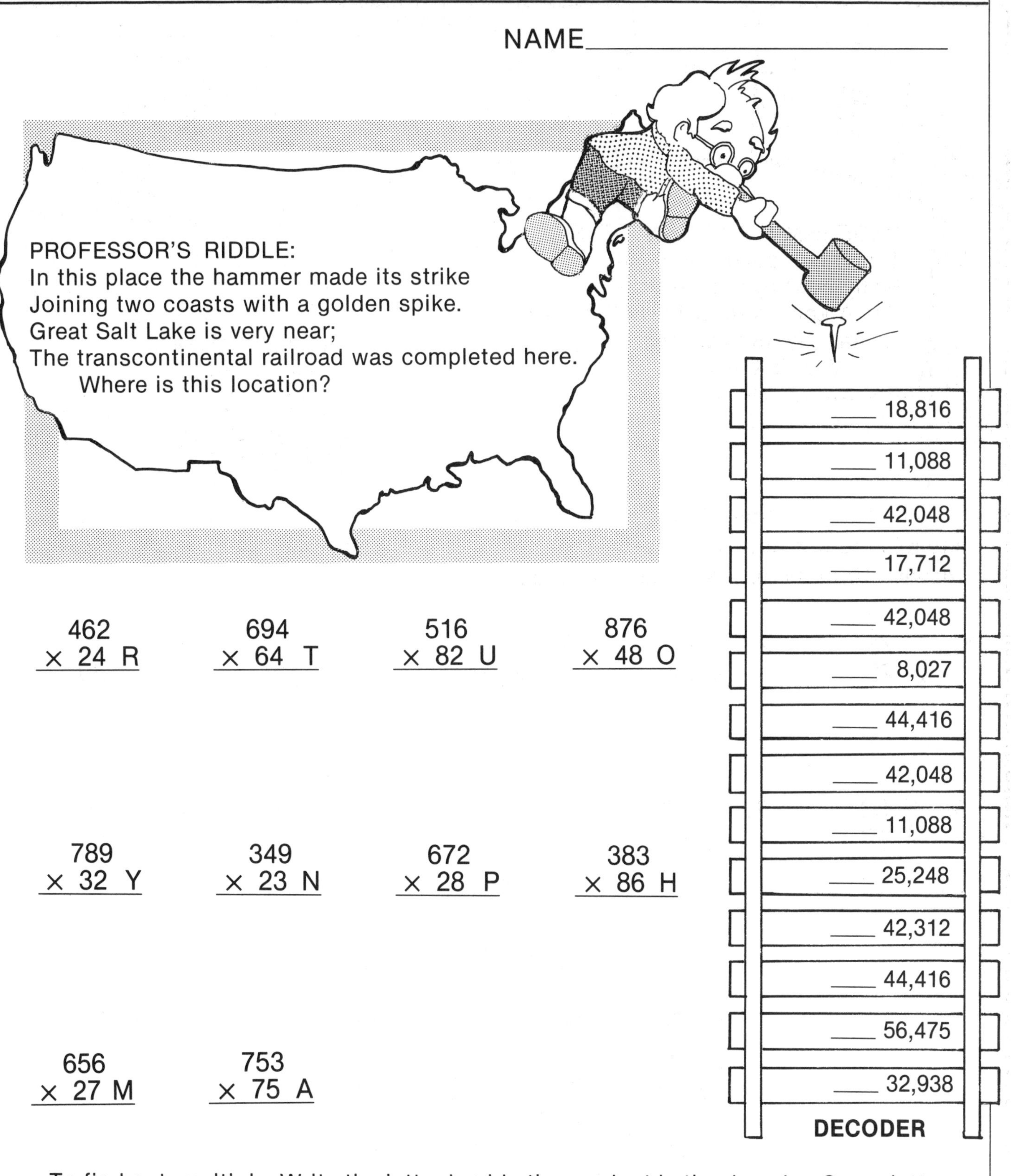

462×24 R

694×64 T

516×82 U

876×48 O

789×32 Y

349×23 N

672×28 P

383×86 H

656×27 M

753×75 A

DECODER

___ 18,816
___ 11,088
___ 42,048
___ 17,712
___ 42,048
___ 8,027
___ 44,416
___ 42,048
___ 11,088
___ 25,248
___ 42,312
___ 44,416
___ 56,475
___ 32,938

To find out, multiply. Write the letter beside the product in the decoder. Some letters are used more than once. The letters spell the answer. Write the letters in the blanks below.

ANSWER: ___ ___ ___ ___ ___ ___ ___ ___ ___ ___,

___ ___ ___ ___

NAME________________________

Hello Friends,

Minnie Places and I visited the city near the geographic center of North America. While we sat in an old covered wagon, a photographer took our picture. The memories of days gone by suddenly came alive. Do you know where we were touring?

Professor Ben Around

To find out, multiply. Write the matching letter in the blank above the product. The letters spell the answer. Some letters will not be used.

ANSWER: ____ ____ ____ ____ ____, ____ ____

392,291 141,453 222,598 403,354 206,416 785,862 637,056

$5{,}432 \times 38$ Y

$6{,}123 \times 59$ O

$9{,}138 \times 53$ H

$5{,}239 \times 27$ U

$7{,}635 \times 98$ T

$6{,}649 \times 59$ R

$4{,}586 \times 66$ E

$7{,}938 \times 99$ N

$6{,}547 \times 34$ G

$8{,}064 \times 79$ D

$5{,}466 \times 93$ S

$8{,}582 \times 47$ B

NAME________________________

PROFESSOR'S MAP MYSTERY:
Down the Snake River kayakers ride;
Canyons running deep and wide.
Potatoes are its claim to fame,
Boiled or baked.
What is this state's name?

To find out, multiply. Write the matching letter in the blank above the product. The letters spell the answer. Some letters will not be used.

ANSWER: ___ ___ ___ ___ ___

___	___	___	___	___
886,121	2,958,039	2,386,524	3,484,167	4,067,514

$62{,}937 \times 47$ D

$70{,}395 \times 88$ E

$95{,}856 \times 78$ W

$75{,}786 \times 65$ M

$65{,}739 \times 53$ H

$85{,}233 \times 28$ A

$41{,}086 \times 99$ O

$54{,}932 \times 87$ S

$64{,}264 \times 39$ N

$38{,}527 \times 23$ I

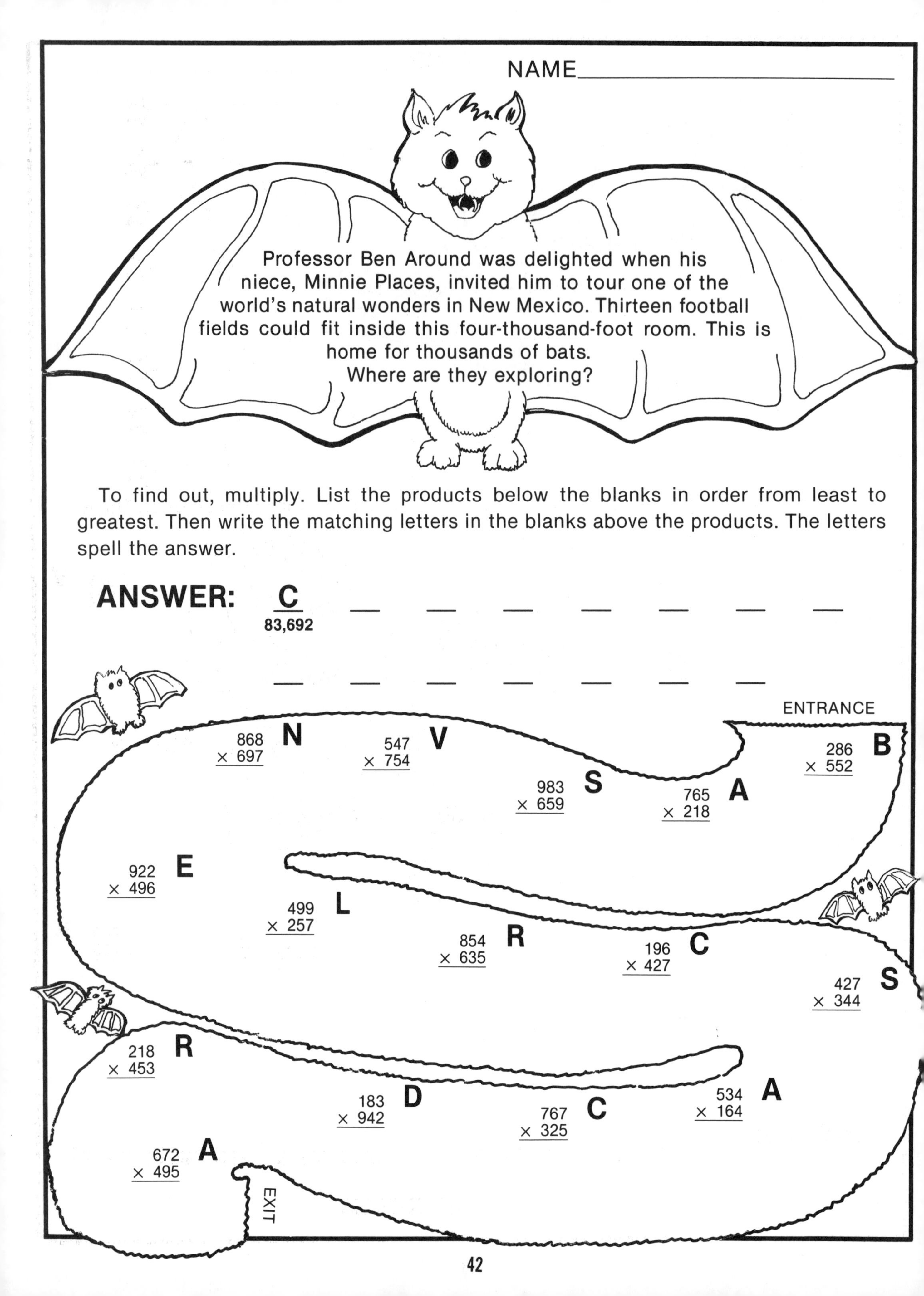

NAME________________________

Professor Ben Around was delighted when his niece, Minnie Places, invited him to tour one of the world's natural wonders in New Mexico. Thirteen football fields could fit inside this four-thousand-foot room. This is home for thousands of bats.
Where are they exploring?

To find out, multiply. List the products below the blanks in order from least to greatest. Then write the matching letters in the blanks above the products. The letters spell the answer.

ANSWER: C (83,692) __ __ __ __ __ __ __

__ __ __ __ __ __ __

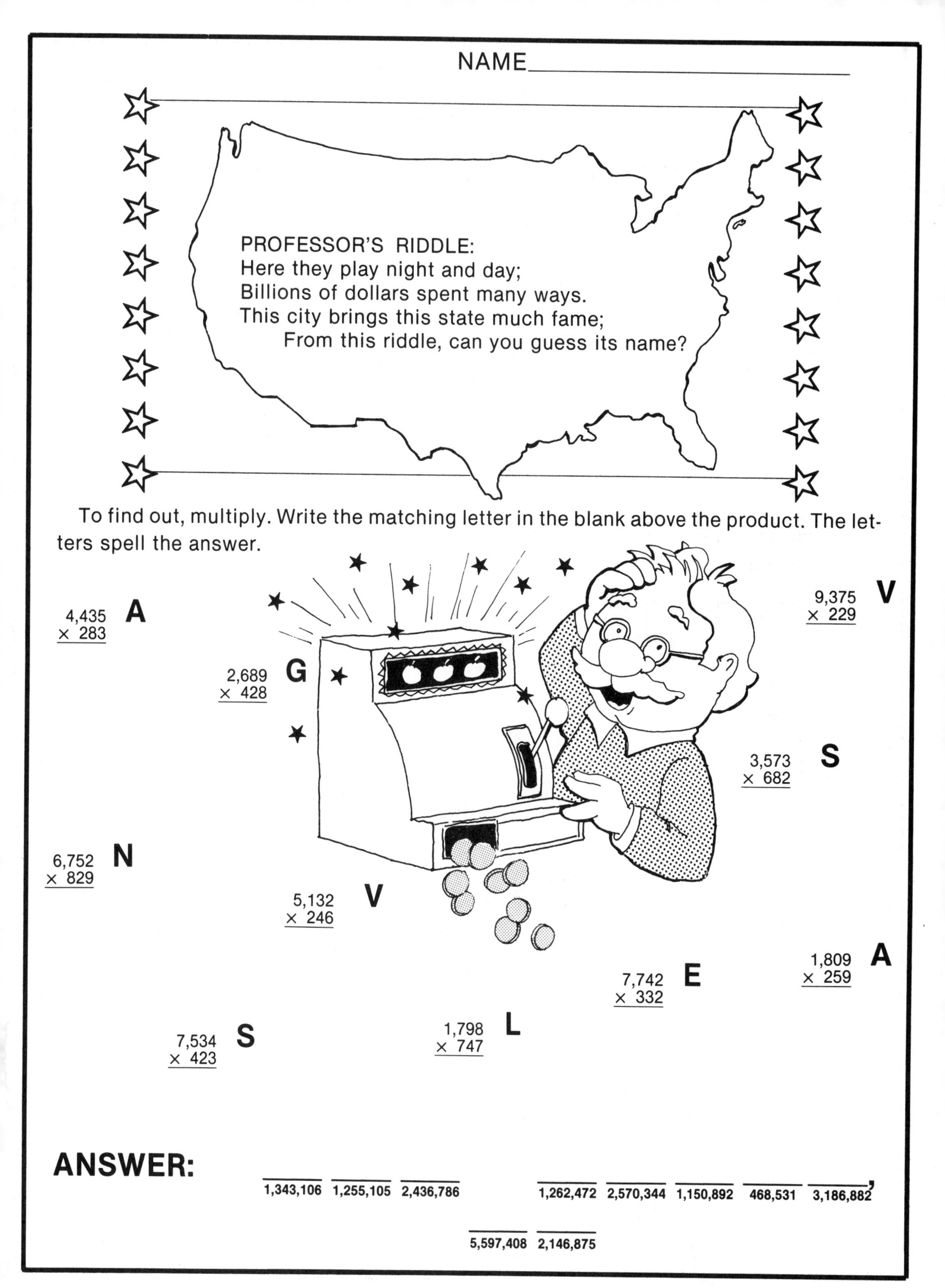

NAME________________________________

PROFESSOR'S RIDDLE:
Here they play night and day;
Billions of dollars spent many ways.
This city brings this state much fame;
From this riddle, can you guess its name?

To find out, multiply. Write the matching letter in the blank above the product. The letters spell the answer.

$$\begin{array}{r} 4{,}435 \\ \times\ 283 \\ \hline \end{array}$$ **A**

$$\begin{array}{r} 9{,}375 \\ \times\ 229 \\ \hline \end{array}$$ **V**

$$\begin{array}{r} 2{,}689 \\ \times\ 428 \\ \hline \end{array}$$ **G**

$$\begin{array}{r} 3{,}573 \\ \times\ 682 \\ \hline \end{array}$$ **S**

$$\begin{array}{r} 6{,}752 \\ \times\ 829 \\ \hline \end{array}$$ **N**

$$\begin{array}{r} 5{,}132 \\ \times\ 246 \\ \hline \end{array}$$ **V**

$$\begin{array}{r} 1{,}809 \\ \times\ 259 \\ \hline \end{array}$$ **A**

$$\begin{array}{r} 7{,}742 \\ \times\ 332 \\ \hline \end{array}$$ **E**

$$\begin{array}{r} 1{,}798 \\ \times\ 747 \\ \hline \end{array}$$ **L**

$$\begin{array}{r} 7{,}534 \\ \times\ 423 \\ \hline \end{array}$$ **S**

ANSWER:

____ ____ ____   ____ ____ ____ ____ ____,
1,343,106 1,255,105 2,436,786   1,262,472 2,570,344 1,150,892 468,531 3,186,882

____ ____
5,597,408 2,146,875

NAME ____________________

Professor Ben Around had an eerie feeling as he walked the streets of an old abandoned town. When Barksalot, his dog, tucked his tail and ran, the professor saw three ghosts walking hand in hand. What state is he visiting?

To find out, multiply. Write the matching letter in the blank above the product. The letters spell the answer.

HINT: An exponent tells how many times a number is used as a factor.

ANSWER:

____	____	____	____	____	____	____
625	4,096	1,000,000	1,000	1,296	64	343

N 10^6 = ______

O 4^6 = ______

A 7^3 = ______

A 6^4 = ______

N 2^6 = ______

M 5^4 = ______

T 10^3 = ______

NAME____________________________

PROFESSOR'S MAP MYSTERY:
The pursuit of learning is important here;
Only the best students come each year.
Yale University is older than most;
This city serves as its host.
What is this city's name?

To find out, multiply. Write the matching letter in the blank above the product. The letters spell the answer.

ANSWER:

___	___	___		___	___	___	___	___,
11,313,820	22,131,347	6,380,850		25,292,251	20,089,342	41,124,240	18,792,690	36,197,402

___	___
25,666,055	3,233,164

H	E	T	E	N
$5{,}123 \times 4{,}937$	$4{,}926 \times 3{,}815$	$2{,}167 \times 1{,}492$	$5{,}789 \times 3{,}823$	$7{,}385 \times 1{,}532$

V	C	W	A	N
$8{,}532 \times 4{,}820$	$5{,}795 \times 4{,}429$	$2{,}575 \times 2{,}478$	$5{,}134 \times 3{,}913$	$6{,}046 \times 5{,}987$

NAME______________________________

Dear Boys and Girls,

Uncle GoGo and I feel like true spelunkers! We're exploring the longest cave in the world. There are more than 200 miles of underground passages. What is the name of this cave in the Bluegrass State?

Professor Ben Around

To find out, complete each division fact. Cross out the space in the decoder that contains each answer. The letters in the remaining spaces from left to right spell the name of the cave. Write the letters in the blanks below.

DECODER

D 72	R 3	C 49	M 51	B 8	A 17	R 7	
G 24	L 16	M 22	S 15	C 9	M 81	B 4	
I 6	O 23	E 12	A 56	T 31	D 5	H 29	
T 13	E 2	K 36	C 26	A 10	V 48	O 11	
E 1	R 40	A 43	N 28	S 63	V 57	E 61	

1. 42 ÷ 6 = ______
2. 32 ÷ 4 = ______
3. ______ ÷ 7 = 8
4. 24 ÷ ______ = 8
5. 18 ÷ ______ = 3
6. ______ ÷ 7 = 4
7. ______ ÷ 9 = 8
8. 25 ÷ 5 = ______
9. 44 ÷ 4 = ______
10. ______ ÷ 7 = 7
11. ______ ÷ 5 = 8
12. 63 ÷ 7 = ______
13. 16 ÷ 4 = ______
14. ______ ÷ 6 = 4
15. 14 ÷ 7 = ______
16. 30 ÷ 3 = ______
17. ______ ÷ 3 = 5
18. ______ ÷ 2 = 6
19. ______ ÷ 6 = 6
20. 14 ÷ ______ = 14
21. 39 ÷ ______ = 3
22. ______ ÷ 9 = 7
23. 16 ÷ ______ = 1
24. ______ ÷ 6 = 8

ANSWER: __ __ __ __ __ __ __   __ __ __ __

NAME________________________

PROFESSOR'S RIDDLE:
The deepest lake
In the USA
In an Oregon volcano,
It rests today.
What is its name?

To find out, divide. Write the matching letter in the blank above the quotient. Some letters will not be used.

B. $7\overline{)25}$	C. $6\overline{)35}$	O. $5\overline{)29}$	R. $8\overline{)41}$
A. $9\overline{)55}$	T. $5\overline{)32}$	S. $6\overline{)55}$	E. $9\overline{)24}$
R. $5\overline{)33}$	Y. $9\overline{)39}$	L. $6\overline{)43}$	W. $7\overline{)41}$
A. $7\overline{)38}$	P. $8\overline{)53}$	K. $3\overline{)25}$	E. $4\overline{)22}$

ANSWER:

____ ____ ____ ____ ____ ____   ____ ____ ____ ____

5 R5   5 R1   6 R1   6 R2   2 R6   6 R3   7 R1   5 R3   8 R1   5 R2

NAME________________________________

While fishing in the largest tributary of the Ohio River, Professor Ben Around and his dog, Barksalot, watched a riverboat go by. Barksalot got so excited he frightened away the fish! In what body of water was the professor fishing?

To find out, divide. Cross out the letter beside each quotient in the decoder. The remaining letters spell the answer. Write the letters in the blanks below.

1.	$5\overline{)89}$	2.	$4\overline{)72}$	3.	$7\overline{)83}$
4.	$5\overline{)64}$	5.	$5\overline{)87}$	6.	$3\overline{)58}$
7.	$4\overline{)62}$	8.	$5\overline{)73}$	9.	$4\overline{)64}$
10.	$7\overline{)82}$	11.	$3\overline{)64}$	12.	$5\overline{)85}$
13.	$4\overline{)50}$	14.	$3\overline{)46}$	15.	$5\overline{)60}$

DECODER

M	12 R2
I	11 R6
T	23
S	17
S	17 R2
I	15 R1
E	22 R5
S	17 R4
N	16 R7
S	12
I	16
N	9 R1
E	24
P	12 R4
P	14 R3
S	16 R8
S	12 R1
I	11 R5
E	27
E	32
A	15 R2
R	19 R6
N	18
I	30 R4
S	19 R1
V	13 R5
E	20
R	34
A	21 R1

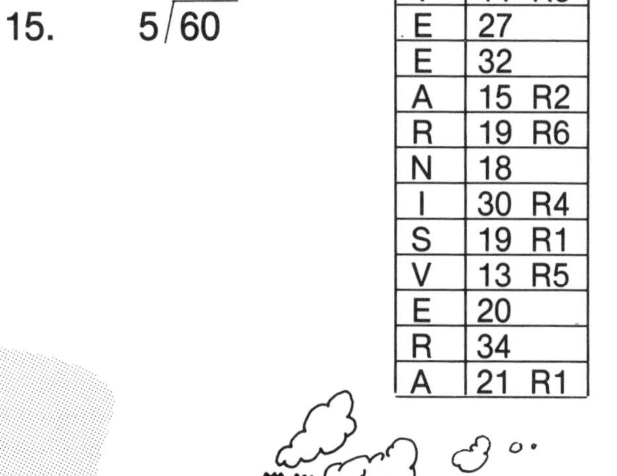

ANSWER: _ _ _ _ _ _ _ _ _ _ _ _ _   _ _ _ _ _ _ _

NAME________________________________

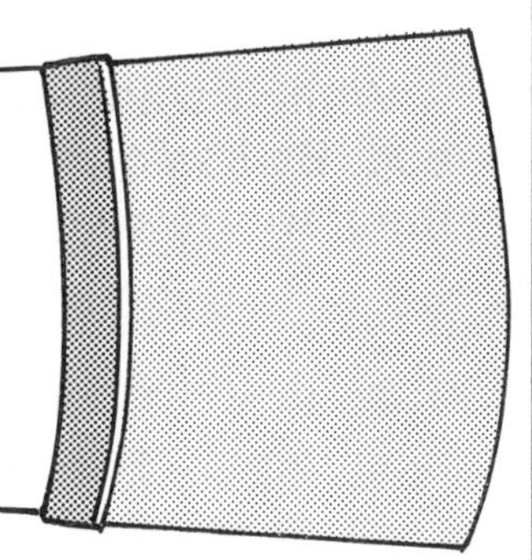

PROFESSOR'S MAP MYSTERY: Professor Ben Around loves discovering the wonders of the night sky as he looks through the biggest telescope in the United States. Where is he?

To find out, divide. Use the star decoder to find the letter that matches each quotient. Write the letter in the blank above the number of the problem. The letters spell the answer.

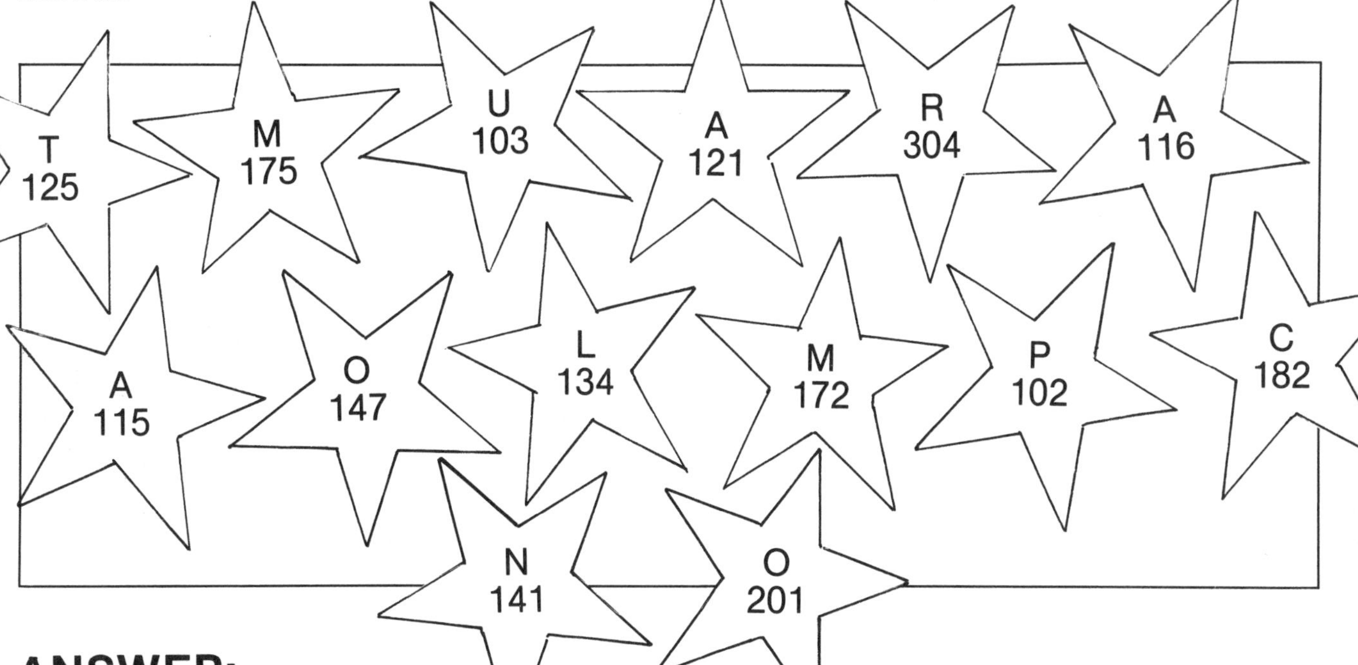

ANSWER:

___ ___ ___ ___ ___   ___ ___ ___ ___ ___ ___ ___,   ___ ___

1 2 3 4 5   6 7 8 9 10 11 12   13 14

1. $5\overline{)860}$ 2. $4\overline{)804}$ 3. $5\overline{)515}$ 4. $3\overline{)423}$

5. $4\overline{)500}$ 6. $6\overline{)612}$ 7. $8\overline{)928}$ 8. $4\overline{)536}$

9. $5\overline{)735}$ 10. $3\overline{)525}$ 11. $8\overline{)920}$ 12. $3\overline{)912}$

13. $5\overline{)910}$ 14. $6\overline{)726}$

NAME ______________________

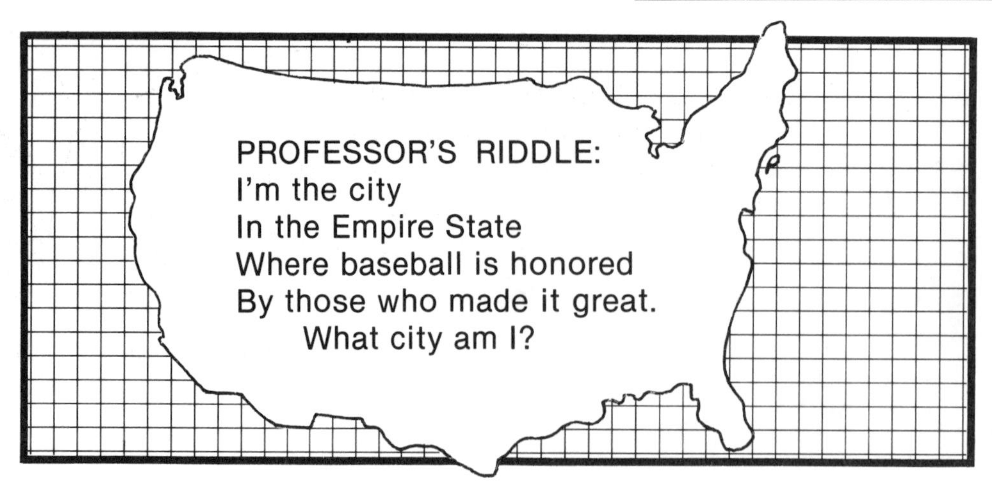

To find out, divide. Cross out the letter beside each quotient in the decoder. The remaining letters spell the answer. Write the letters in the blanks below.

1. $4\overline{)812}$ 2. $6\overline{)636}$ 3. $6\overline{)3024}$ 4. $5\overline{)3010}$

5. $8\overline{)872}$ 6. $2\overline{)808}$ 7. $2\overline{)1414}$ 8. $6\overline{)3630}$

9. $4\overline{)2812}$ 10. $9\overline{)3609}$ 11. $5\overline{)2540}$

12. $3\overline{)1809}$ 13. $7\overline{)5614}$ 14. $6\overline{)4236}$

15. $8\overline{)4056}$ 16. $4\overline{)2404}$

DECODER

B	707
C	102
R	508
O	205
D	601
O	308
K	605
P	409
L	507
E	504
T	603
E	107
Y	401
N	706
R	501
O	109
I	404
S	208
T	406
F	703
O	304
A	802
W	608
N	306
M	106
N	902
Y	408
I	203
A	602

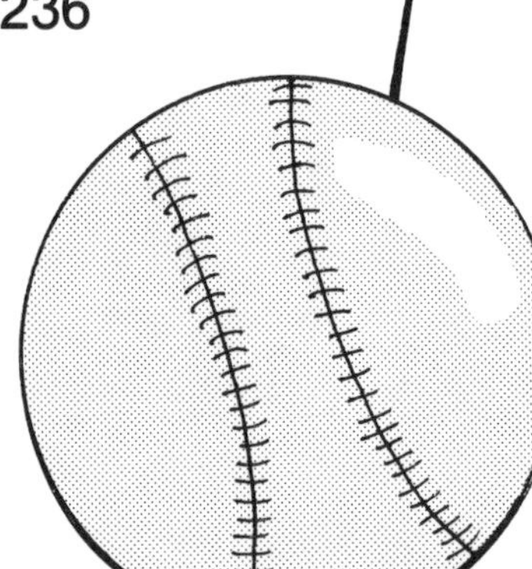

ANSWER: _ _ _ _ _ _ _ _ _ _ _ _ _ _ _ _, _ _

NAME________________________________

PROFESSOR'S MAP MYSTERY:
Birds, mammals, and reptiles, too;
The largest animal collection is in this zoo.
Where is this zoo located?

To find out, divide. Write the matching letter in the blank above the number of the problem. The letters spell the answer.

ANSWER:

__ __ __   __ __ __ __ __,   __ __ __ __ __ __ __ __ __ __

1 2 3   4 5 6 7 8   9 10 11 12 13 14 15 16 17 18

104	201	30	71	91	132	21	120
R	L	A	D	G	O	C	A

108	93	80	52	50	31	105	109	103	62
N	I	I	E	O	S	I	A	F	N

1. $8\overline{)248}$
2. $5\overline{)150}$
3. $3\overline{)186}$
4. $2\overline{)142}$
5. $7\overline{)560}$
6. $4\overline{)208}$
7. $9\overline{)819}$
8. $5\overline{)250}$
9. $6\overline{)126}$
10. $3\overline{)360}$
11. $4\overline{)804}$
12. $5\overline{)525}$
13. $7\overline{)721}$
14. $2\overline{)264}$
15. $8\overline{)832}$
16. $9\overline{)972}$
17. $3\overline{)279}$
18. $4\overline{)436}$

NAME________________________________

Professor Ben Around is browsing through the oldest library in the United States. Benjamin Franklin founded the library in 1731. What city is the professor visiting?

To find out, divide. Write the matching letter in the blank above the number of the problem. The letters spell the answer.

ANSWER: __ __ __ __ __ __ __ __ __ __ __ __, __ __

1 2 3 4 5 6 7 8 9 10 11 12 13 14

1. $5\overline{)612}$ 2. $3\overline{)704}$ 3. $5\overline{)294}$ 4. $4\overline{)631}$

5. $7\overline{)304}$ 6. $5\overline{)673}$ 7. $4\overline{)594}$ 8. $3\overline{)269}$

9. $4\overline{)207}$ 10. $8\overline{)310}$ 11. $7\overline{)937}$

12. $8\overline{)825}$ 13. $9\overline{)530}$ 14. $6\overline{)435}$

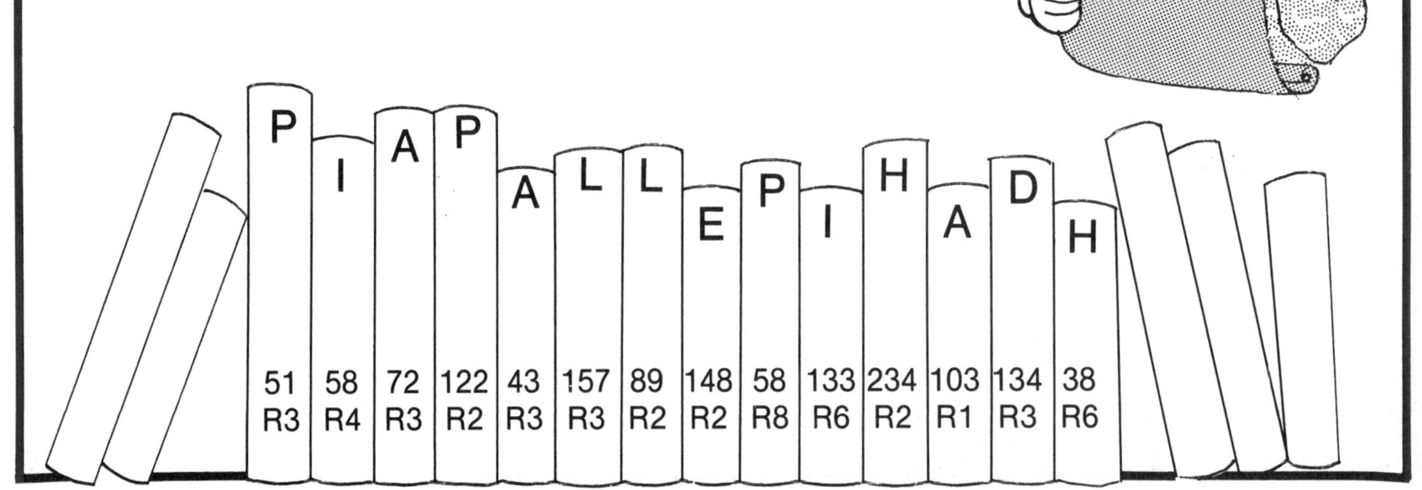

NAME ____________________

PROFESSOR'S RIDDLE:
Along the coast Blackbeard came;
Piracy was the name of his game.
Gold, jewels, and treasures he stole;
For this his head was put on a pole.
Near what coast was Blackbeard executed?

ANSWER:

_ _ _ _ _ _

_ _ _ _ _ _ _ _ _ _

To find out, divide. Draw a path connecting the quotients in order of the problems. The letters along the path spell the answer. Write the letters in the blanks below.

1. $21\overline{)7643}$
2. $34\overline{)8156}$
3. $44\overline{)3417}$
4. $52\overline{)5816}$
5. $67\overline{)8245}$
6. $41\overline{)3986}$
7. $25\overline{)1982}$
8. $83\overline{)2350}$
9. $46\overline{)5083}$
10. $59\overline{)4031}$
11. $61\overline{)3215}$
12. $29\overline{)4526}$
13. $76\overline{)6588}$

START	363 R20 N	238 R14 I	79 R7 A	28 R26 R
362 R16 S	239 R30 O	77 R29 R	97 R9 C	110 R23 O
246 R5 N	76 R31 U	111 R44 T	123 R4 H	68 R19 L
67 R10 E	101 R64 W	86 R52 A	156 R2 N	52 R43 I

NAME________________________

Hello Boys and Girls,

What excitement this city must have known during the gold rush of 1849! Thousands of fortune seekers came by foot, horseback, wagon, and boat to the place that five years later became the capital of the Golden State. What is this city's name?

Professor Ben Around

To find out, divide. Write the matching letter in the blank above the quotient. Some letters will not be used. The letters spell the answer.

ANSWER:

___ ___ ___ ___ ___ ___ ___ ___ ___ ___, ___ ___

14 R6 10 R20 7 R13 13 R4 11 R14 5 R5 7 R14 6 R16 6 R24 8 R16 18 R12 12 R6

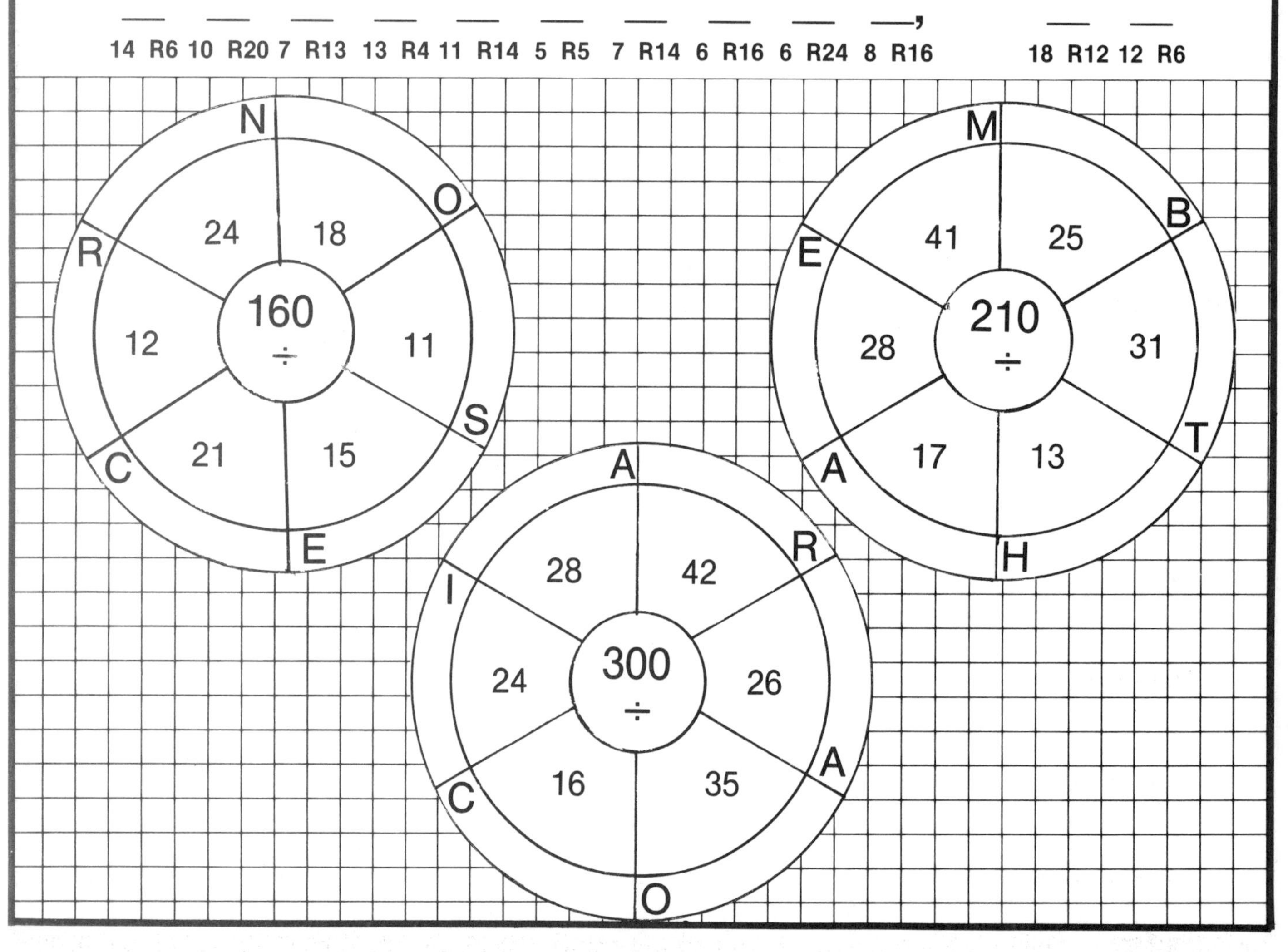

NAME___________________________

PROFESSOR'S MAP MYSTERY:
Salmon jumping up fish ladders
On their way upstream;
A major dam on the Columbia River
Is where they can be seen.
What is this place?

To find out, divide. Draw a line connecting the quotients in order of the problems to discover the outline of the state. Then write the matching letter in the blank above each quotient. The letters spell the answer.

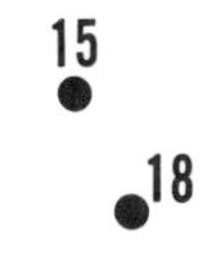

6 12 2 16 7 9 8 4 3 11

ANSWER: ___ ___ ___ ___ ___ ___ ___ ___ ___ ___   ___ ___ ___,   ___ ___

16 8 5 22 18 9 4 15 12 2   6 19 11,   3 7

1. N 205 ÷ 41 = ______	2. I 252 ÷ 63 = ______	3. O 168 ÷ 56 = ______
4. M 704 ÷ 64 = ______	5. O 592 ÷ 74 = ______	6. V 315 ÷ 35 = ______
7. R 644 ÷ 92 = ______	8. B 768 ÷ 48 = ______	9. E 138 ÷ 69 = ______
10. L 348 ÷ 29 = ______	11. D 348 ÷ 58 = ______	12. A 646 ÷ 34 = ______
13. E 936 ÷ 52 = ______	14. L 945 ÷ 63 = ______	15. N 858 ÷ 39 = ______

NAME________________________________

To find out, divide. Write the matching letter in the blank above the quotient. The letters spell the answer. Some letters will not be used.

ANSWER: __ __ __ __ __ __ __   __ __ __ __ __,   __ __

63 65 24 18 95 17 42   98 37 23 69 71'   38 26

R	30)510	L	40)920	F	10)980	G	50)900
Y	70)1820	A	60)2520	N	20)1260	S	80)5680
T	20)1700	I	90)5850	B	50)3400	O	30)1470
A	80)7600	L	70)4830	E	30)1860	A	50)1200
C	60)1680	A	40)1480	H	30)1530	N	70)2660

NAME________________________

The last time Professor Ben Around visited Shenandoah National Park in Virginia, nephew Ben There took a picture of him on Virginia's highest mountain. Although the mountain is only slightly over a mile above sea level, Professor Ben felt just like he was on top of the world! Where were they?

To find out, divide. If the quotient appears in the column of numbers at the right, circle the letter beside the problem. The circled letters spell the answer. Write the letters in the blanks below.

M	17)7004	T	24)8856	C	42)5712
O	16)8048	O	35)7315	U	56)5768
K	28)6916	N	26)7722	T	38)9728
S	41)4346	R	39)5772	B	27)6669
A	43)5289	O	22)5148	T	81)8586
G	72)7272	E	63)9954	Y	54)7938
R	46)8832	S	68)7752		

148
297
450
103
530
412
125
503
518
158
234
101
216
482
114
256
245
192
164
346

ANSWER: _ _ _ _ _ _   _ _ _ _ _ _ _ _

NAME____________________

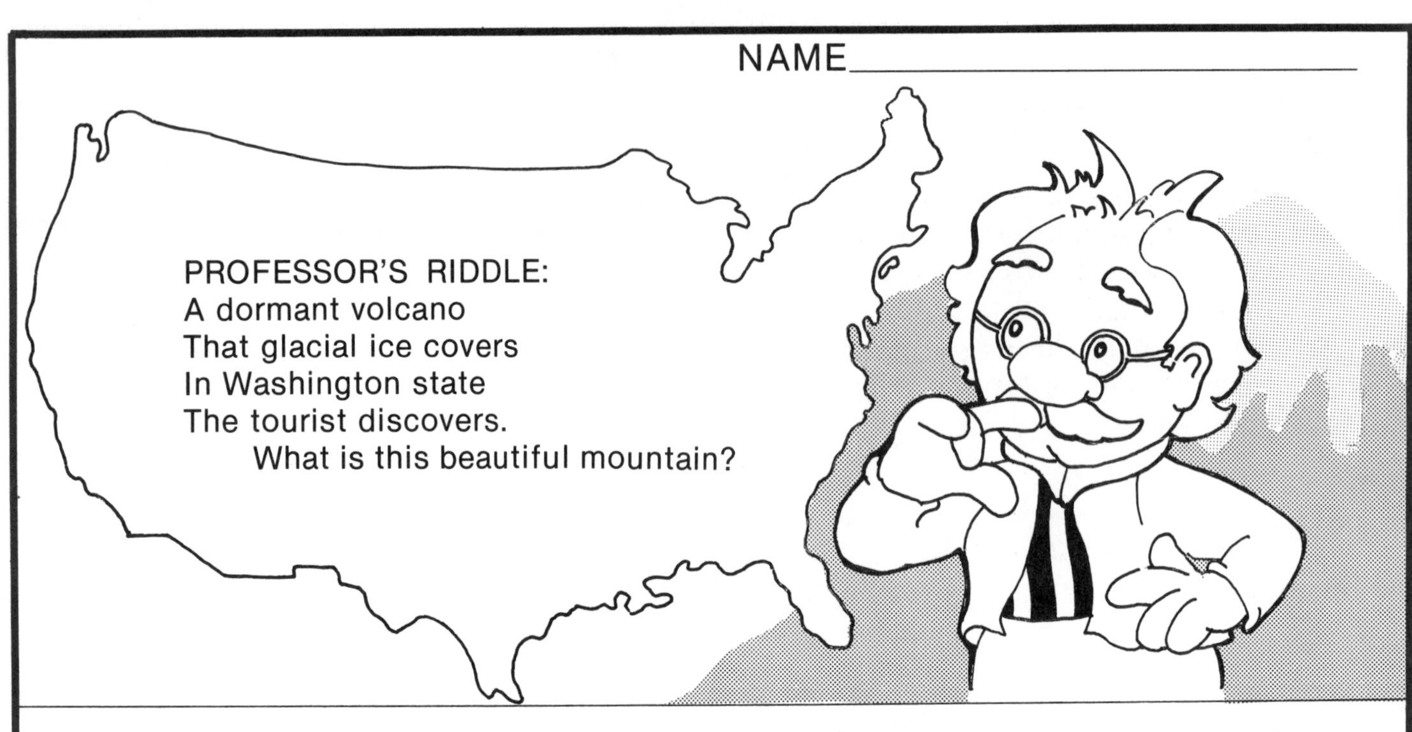

To find out, divide. Write the matching letter in the blank above the quotient. The letters spell the answer.

ANSWER:

__	__	__	__	__		__	__	__	__	__	__	__
93	**61**	**19**	**32**	**34**		**78**	**50**	**43**	**27**	**58**	**25**	**83**

N $406\overline{)12{,}992}$ I $394\overline{)22{,}852}$ O $561\overline{)34{,}221}$

R $732\overline{)57{,}096}$ M $205\overline{)19{,}065}$ E $483\overline{)12{,}075}$

I $534\overline{)22{,}962}$ A $253\overline{)12{,}650}$ R $173\overline{)14{,}359}$

N $560\overline{)15{,}120}$ U $591\overline{)11{,}229}$ T $367\overline{)12{,}478}$

NAME ____________________

To find out, divide. Cross out the letter above each quotient in the decoder. The remaining letters spell the answer. Write the letters in the blanks below.

DECODER

D	V	M	E	I	L	R	G	A	E
19 R4	21 R3	56 R16	28 R4	14 R52	31 R7	16 R20	15 R16	27 R9	12 R2
G	W	I	O	A	N	I	R	A	E
17 R2	26 R4	36 R1	16 R24	39 R4	11 R7	24 R4	22 R5	19 R6	18 R7

1. $214\overline{)7705}$

2. $325\overline{)5220}$

3. $178\overline{)3741}$

4. $482\overline{)5786}$

5. $231\overline{)5548}$

6. $506\overline{)8604}$

7. $354\overline{)6732}$

8. $462\overline{)5089}$

9. $134\overline{)7520}$

10. $473\overline{)7592}$

11. $568\overline{)8004}$

12. $630\overline{)9466}$

ANSWER: __ __ __ __ __ __ __ __

PROFESSOR'S RIDDLE:
The fifth smallest state
In the USA
Has resorts, parks, and beaches
Where the professor likes to play.
What is this state?

NAME________________________

To find out, divide. Draw a line from the letter beside each problem to the quotient. Write the letter in the blank beside the quotient. The letters spell the answer. Write the letters in the blanks below.

Problem	Letter	Quotient	
1. 1,184 ÷ 4 =	N	680	_____
2. 2,208 ÷ 8 =	H	276	_____
3. 1,530 ÷ 5 =	E	695	_____
4. 4,869 ÷ 9 =	F	184	_____
5. 2,085 ÷ 3 =	E	741	_____
6. 4,446 ÷ 6 =	T	453	_____
7. 4,144 ÷ 7 =	S	286	_____
8. 7,800 ÷ 24 =	Y	306	_____
9. 7,020 ÷ 36 =	R	203	_____
10. 7,728 ÷ 42 =	S	541	_____
11. 8,160 ÷ 12 =	T	296	_____
12. 6,293 ÷ 31 =	O	324	_____
13. 7,248 ÷ 16 =	A	486	_____
14. 7,290 ÷ 15 =	W	318	_____
15. 8,100 ÷ 25 =	E	524	_____
16. 9,956 ÷ 19 =	E	195	_____
17. 7,632 ÷ 24 =	J	592	_____
18. 7,293 ÷ 17 =	E	429	_____
19. 6,292 ÷ 22 =	T	325	_____

ANSWER: ___ ___ ___   ___ ___ ___ ___ ___   ___ ___

___ ___ ___   ___ ___ ___ ___ ___ ___

NAME________________________

PROFESSOR'S RIDDLE:
They say Jim Bridger saw it first;
Its water cannot quench your thirst.
A beautiful spot for all to see,
Utah is the place to be.
What is the name of this place?

To find out, draw a line from each fraction to the equivalent part. Write the matching letter in the blank above the number of the problem. The letters spell the answer.

1. 9/14	•	•	E
2. 5/6	•	•	S
3. 7/16	•	•	T
4. 1/2	•	•	G
5. 3/8	•	•	A
6. 11/12	•	•	R
7. 1/3	•	•	L
8. 7/12	•	•	E
9. 1/6	•	•	A
10. 9/16	•	•	T
11. 4/9	•	•	K
12. 4/10	•	•	A
13. 5/8	•	•	L

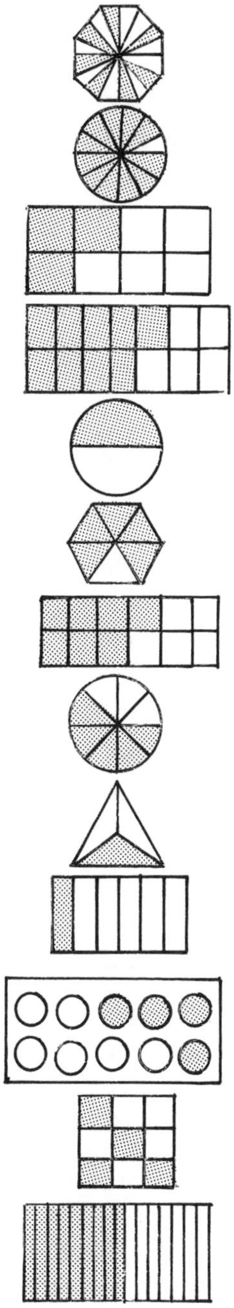

ANSWER:

__ __ __ __ __   __ __ __ __   __ __ __ __

1 2 3 4 5   6 7 8 9   10 11 12 13

NAME________________________

Rise and shine! Professor Ben Around is one of the first Americans to see the sun rise. He is visiting the place where the sun's rays first touch the United States. Remember . . . the sun rises in the east and sets in the west.

Where is he visiting?

To find out, circle the answer below each problem. The letters below each circled answer spell the name of this place. Write the letters in the blanks below.

1. 1/9 of 63 =
   6 8 7
   B P M

2. 1/10 of 40 =
   10 4 40
   C O F

3. 1/4 of 28 =
   7 14 12
   U G V

4. 1/2 of 36 =
   16 18 20
   R N S

5. 5/6 of 30 =
   30 20 25
   L M T

6. 2/3 of 18 =
   10 12 18
   A K E

7. 3/4 of 32 =
   24 27 21
   A Q W

8. 2/5 of 15 =
   10 6 8
   D T N

9. 3/7 of 56 =
   32 21 24
   P S A

10. 7/8 of 32 =
    24 28 21
    J H B

11. 5/10 of 50 =
    25 30 50
    D C K

12. 3/9 of 81 =
    21 18 27
    F A I

13. 4/6 of 18 =
    9 12 24
    T N G

14. 4/5 of 45 =
    36 27 32
    M F H

15. 4/8 of 88 =
    55 44 40
    L E T

ANSWER: _ _ _ _ _ _   _ _ _ _ _ _ _ _ _ _,   _ _

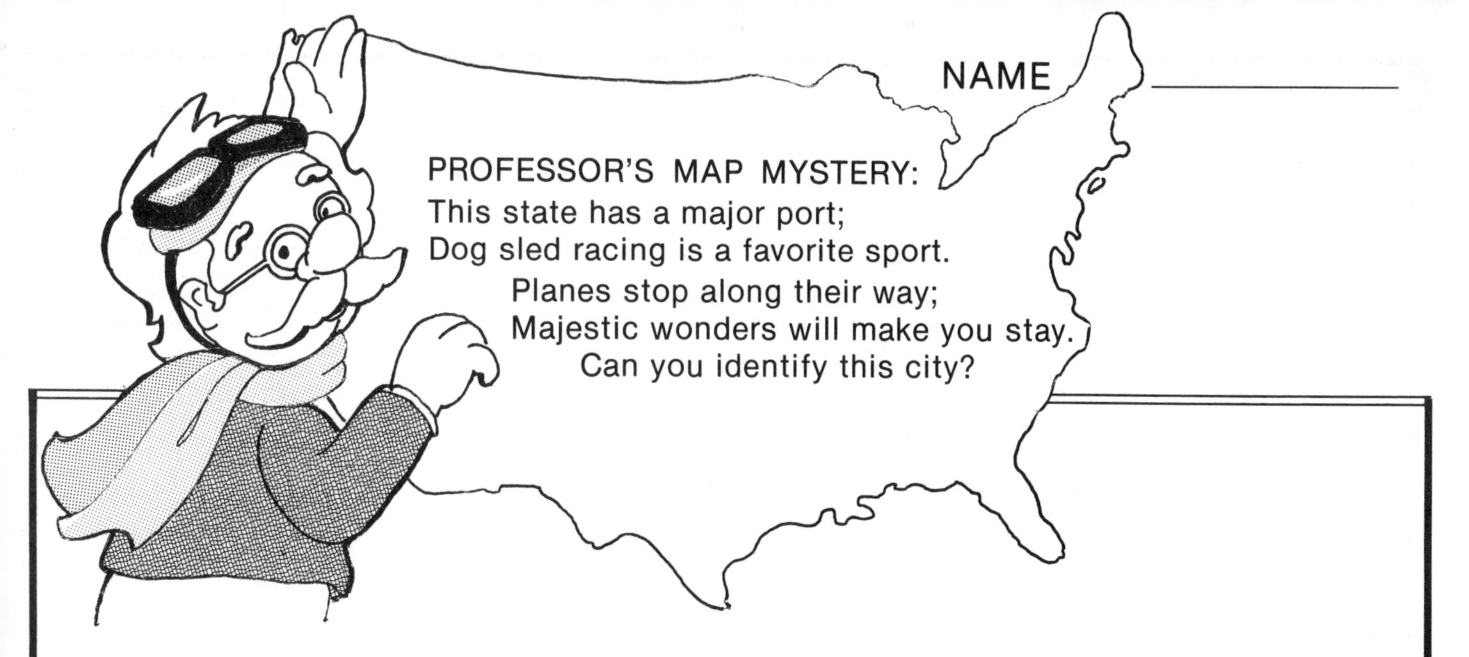

NAME ________________

PROFESSOR'S MAP MYSTERY:
This state has a major port;
Dog sled racing is a favorite sport.
Planes stop along their way;
Majestic wonders will make you stay.
Can you identify this city?

To find out, write the answers in the boxes to make equivalent fractions. In the decoder, begin with the number eighteen and travel clockwise writing the matching letters of the equivalent fractions in the blanks below. The letters spell the answer. Some letters will not be used.

$\frac{1}{3} = \frac{\square}{9}$ **C**	$\frac{3}{4} = \frac{\square}{16}$ **L**	$\frac{5}{7} = \frac{\square}{28}$ **T**	$\frac{1}{6} = \frac{3}{\square}$ **A**
$\frac{1}{4} = \frac{\square}{20}$ **N**	$\frac{1}{5} = \frac{2}{\square}$ **R**	$\frac{3}{5} = \frac{\square}{15}$ **A**	$\frac{1}{2} = \frac{\square}{16}$ **H**
$\frac{1}{4} = \frac{\square}{8}$ **P**	$\frac{3}{4} = \frac{\square}{8}$ **K**	$\frac{1}{3} = \frac{\square}{12}$ **O**	$\frac{2}{8} = \frac{\square}{64}$ **A**
$\frac{3}{4} = \frac{\square}{32}$ **M**	$\frac{5}{7} = \frac{25}{\square}$ **E**	$\frac{5}{6} = \frac{\square}{18}$ **G**	

DECODER

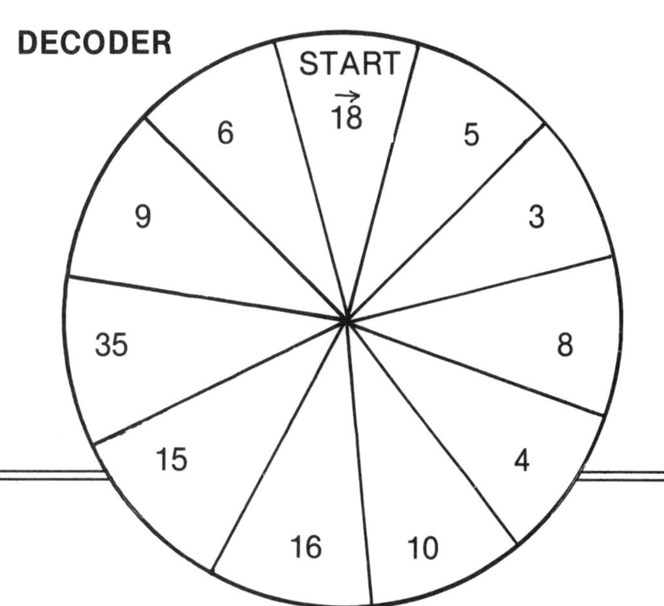

ANSWER: A _ _ _ _ _ _ _ _ _ _ , _ _

NAME____________________

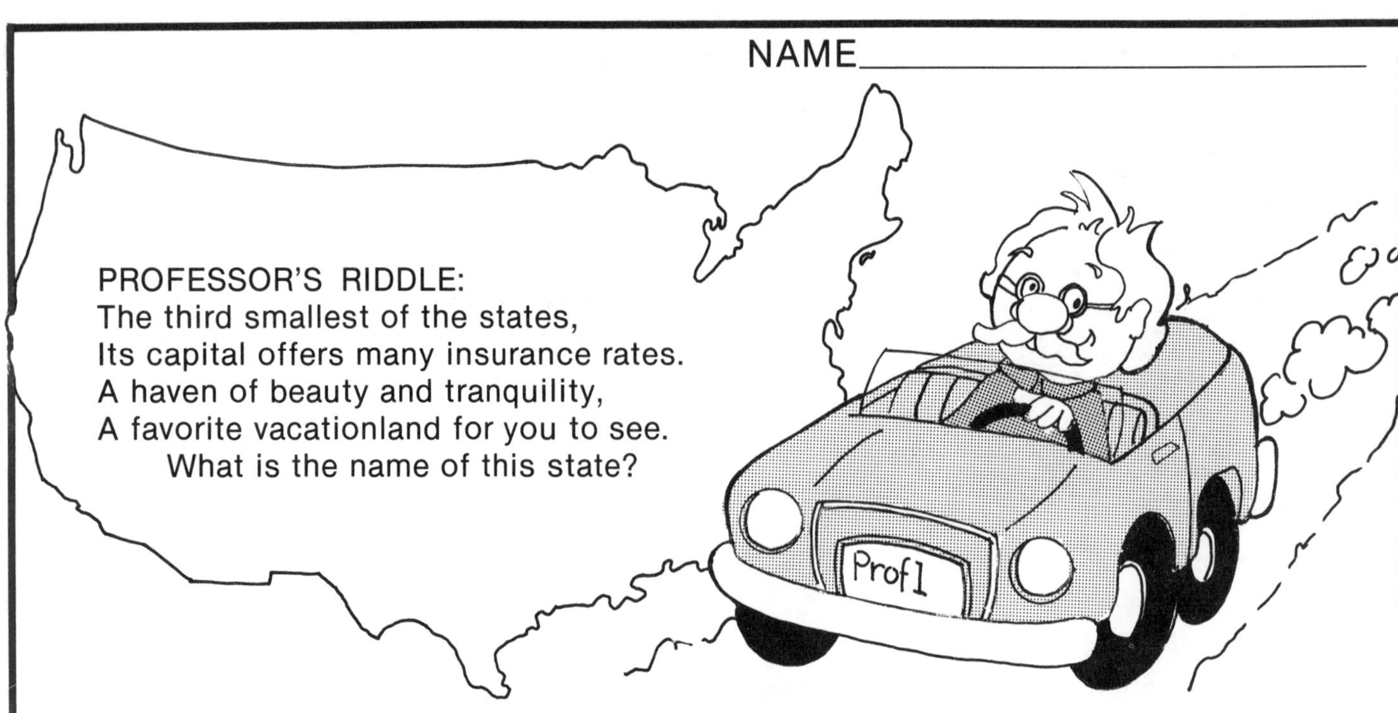

PROFESSOR'S RIDDLE:
The third smallest of the states,
Its capital offers many insurance rates.
A haven of beauty and tranquility,
A favorite vacationland for you to see.
What is the name of this state?

To find out, write each fraction as a whole number or mixed number. Cross out the letter beside each product in the decoder. The remaining letters spell the answer. Write the letters in the blanks below.

13/6 = ____

9/3 = ____

17/7 = ____

13/12 = ____

57/10 = ____

21/4 = ____

7/3 = ____

11/9 = ____

8/2 = ____

5/2 = ____

4/3 = ____

19/6 = ____

27/4 = ____

37/8 = ____

24/5 = ____

18/2 = ____

DECODER

T	3
C	1 2/3
I	5 1/4
D	2 1/3
O	6 2/9
E	6 3/4
U	2 1/2
N	5 1/3
L	2 3/7
N	8 2/7
R	2 1/6
E	3 3/8
W	4 5/8
X	1 1/3
G	4
C	4 3/10
T	10 1/2
A	4 4/5
I	7 3/10
S	5 7/10
C	1 8/9
N	1 1/12
U	3 1/16
H	3 1/6
B	1 2/9
T	6 8/17
K	9

ANSWER: _ _ _ _ _ _ _ _ _ _ _

PROFESSOR'S RIDDLE:
Roger Williams came here first;
Religious freedom was his thirst.
Samuel Slater started a mill;
Textiles this state produces still.
What state is this?

NAME________________________

To find out, reduce. Begin at START and help Professor Ben Around cycle through the maze of reduced fractions by drawing a path in order of the problems. Write the matching letter in the blank above the reduced fraction. The letters spell the answer.

1. 14/22 = ____ 2. 16/18 = ____ 3. 4/6 = ____ 4. 2/10 = ____

5. 9/21 = ____ 6. 4/10 = ____ 7. 6/80 = ____ 8. 16/44 = ____

9. 5/60 = ____ 10. 5/10 = ____ 11. 12/32 = ____ 12. 9/12 = ____

13. 8/14 = ____ 14. 14/18 = ____ 15. 8/10 = ____ 16. 36/42 = ____

17. 40/72 = ____ 18. 6/10 = ____ 19. 4/16 = ____ 20. 2/12 = ____

21. 25/30 = ____

ANSWER:

___ ___ ___   ___ ___ ___ ___ ___   ___ ___
3/40 2/3 4/11   8/9 1/12 1/4 3/5 3/7   6/7 5/9

___ ___ ___ ___ ___   ___ ___ ___ ___ ___ ___
1/5 3/8 4/7 5/6 1/6   4/5 7/11 2/5 3/4 1/2 7/9

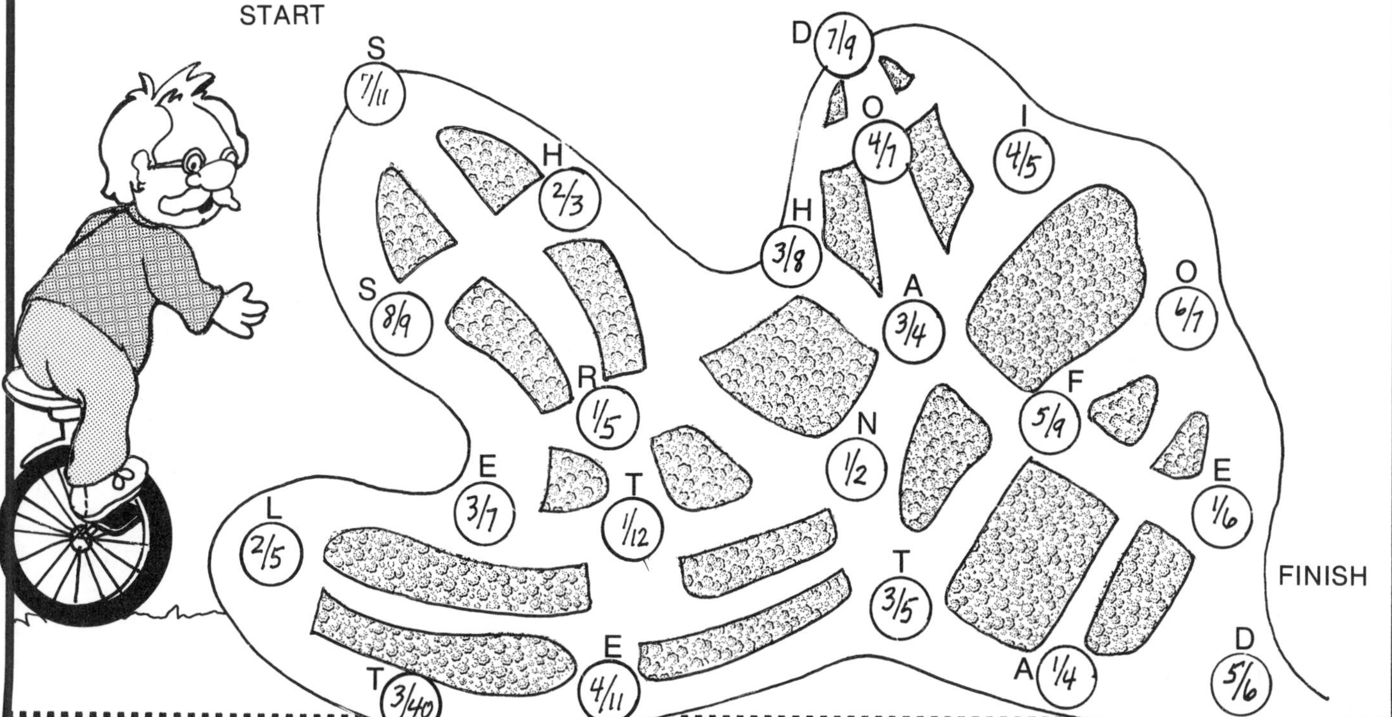

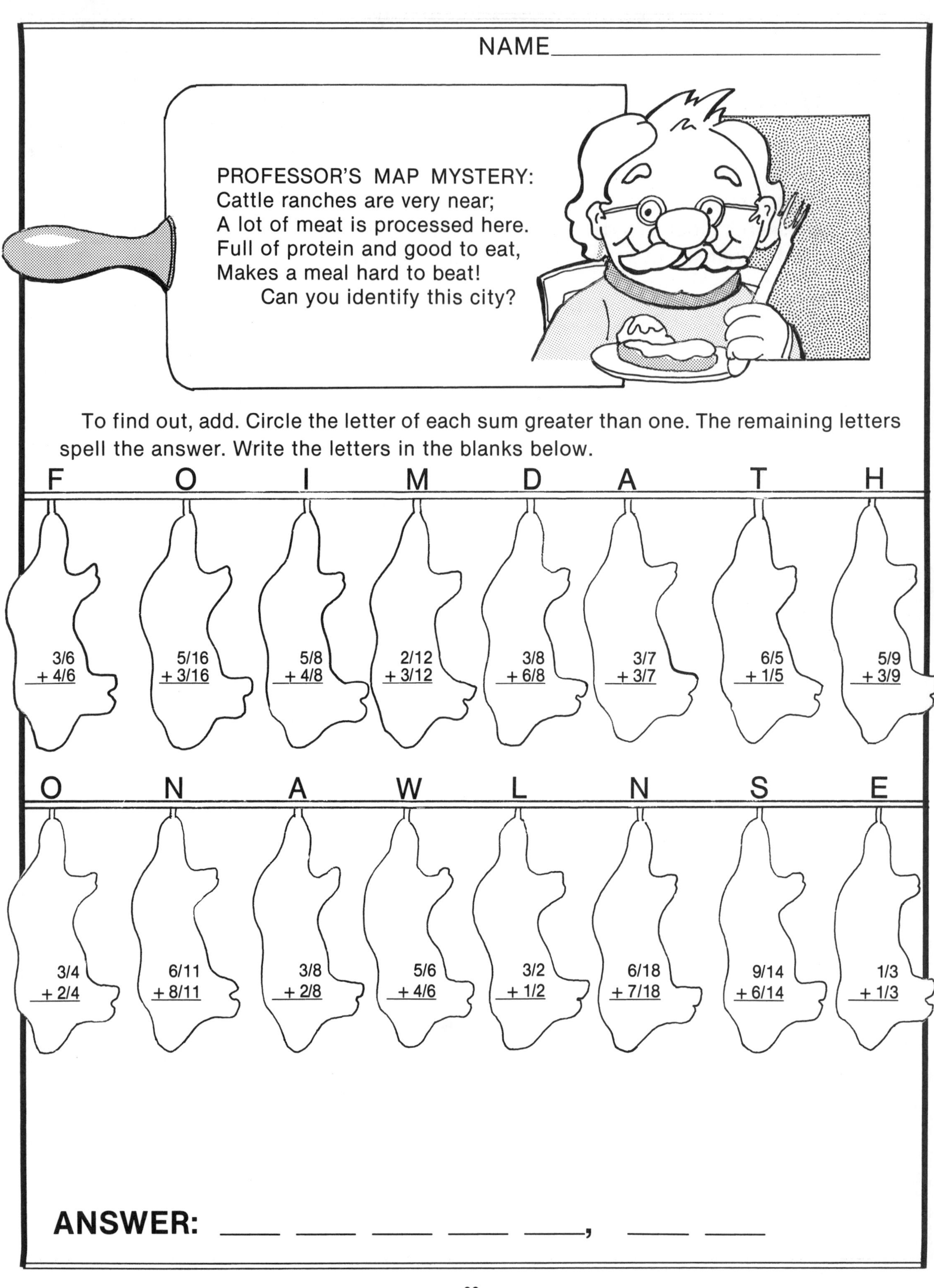
NAME
PROFESSOR'S MAP MYSTERY:
Cattle ranches are very near;
A lot of meat is processed here.
Full of protein and good to eat,
Makes a meal hard to beat!
Can you identify this city?
To find out, add. Circle the letter of each sum greater than one. The remaining letters spell the answer. Write the letters in the blanks below.
F O I M D A T H
3/6 + 4/6
5/16 + 3/16
5/8 + 4/8
2/12 + 3/12
3/8 + 6/8
3/7 + 3/7
6/5 + 1/5
5/9 + 3/9
O N A W L N S E
3/4 + 2/4
6/11 + 8/11
3/8 + 2/8
5/6 + 4/6
3/2 + 1/2
6/18 + 7/18
9/14 + 6/14
1/3 + 1/3
ANSWER: ___ ___ ___ ___ ___, ___ ___

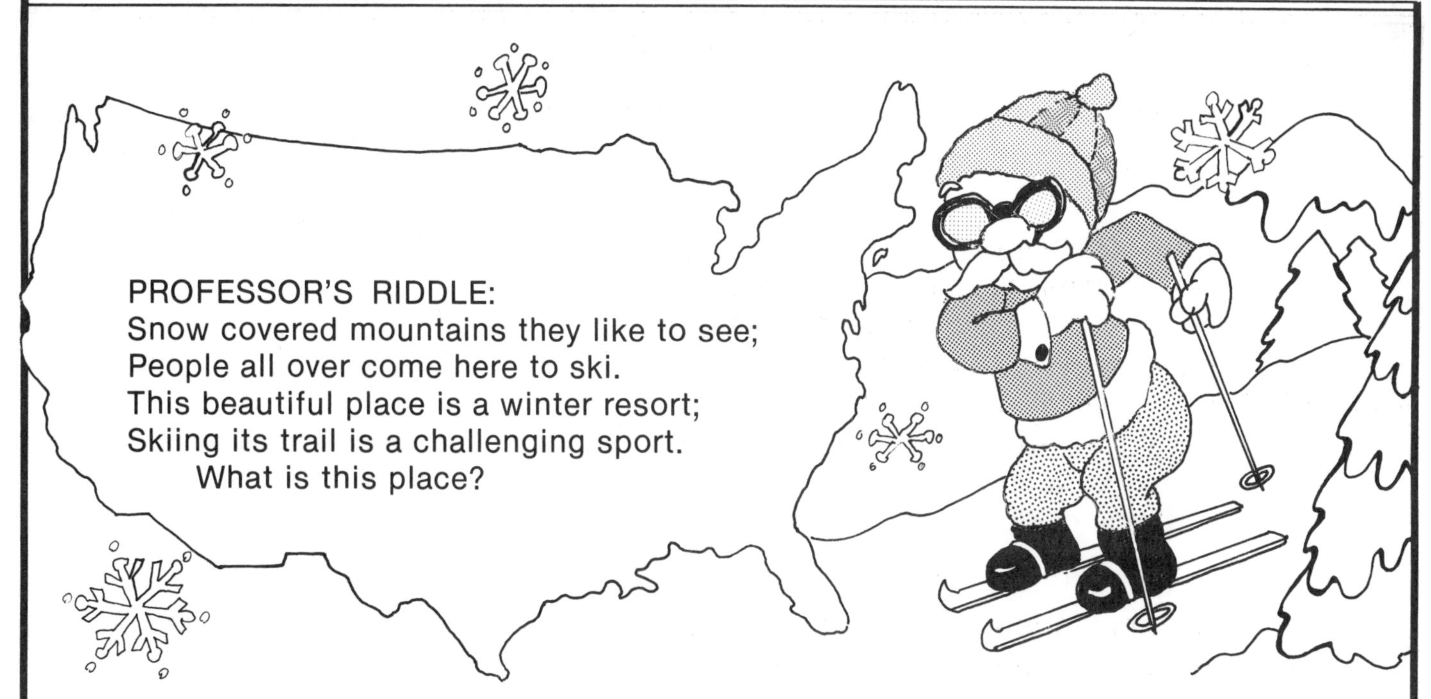

PROFESSOR'S RIDDLE:
Snow covered mountains they like to see;
People all over come here to ski.
This beautiful place is a winter resort;
Skiing its trail is a challenging sport.
What is this place?

To find out, subtract. Write the matching letter in the decoder above the difference. The letters spell the answer. Some letters will not be used.

DECODER

2/9	2/5	2/25	5/9	1/14	1/2	3/17	1/7	1/8	7/18	3/4

2/6 − 1/6 M	4/14 − 3/14 A	4/5 − 3/5 F	12/16 − 4/16 L	8/10 − 2/10 S
8/9 − 3/9 V	7/8 − 1/8 D	4/9 − 2/9 S	20/40 − 15/40 Y	3/12 − 2/12 P
2/7 − 1/7 E	8/13 − 5/13 O	12/20 − 10/20 W	8/10 − 4/10 U	8/15 − 3/15 T
10/17 − 7/17 L	7/14 − 2/14 R	8/25 − 6/25 N	9/11 − 2/11 C	10/18 − 3/18 I

NAME________________________

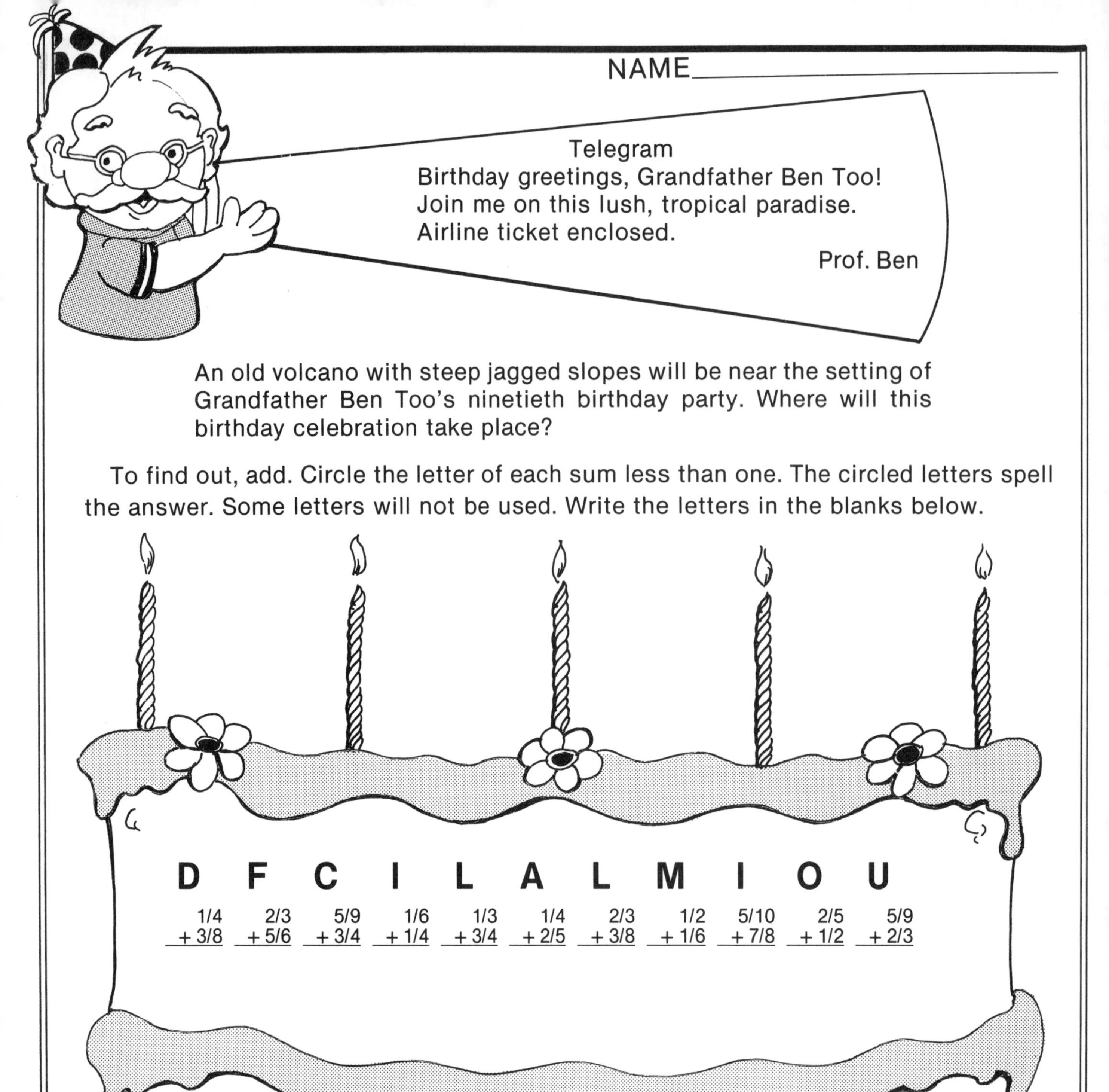

An old volcano with steep jagged slopes will be near the setting of Grandfather Ben Too's ninetieth birthday party. Where will this birthday celebration take place?

To find out, add. Circle the letter of each sum less than one. The circled letters spell the answer. Some letters will not be used. Write the letters in the blanks below.

D	F	C	I	L	A	L	M	I	O	U
1/4	2/3	5/9	1/6	1/3	1/4	2/3	1/2	5/10	2/5	5/9
+ 3/8	+ 5/6	+ 3/4	+ 1/4	+ 3/4	+ 2/5	+ 3/8	+ 1/6	+ 7/8	+ 1/2	+ 2/3

N	D	R	H	I	E	A	G	D	H	I
2/5	1/3	2/3	1/2	3/5	1/3	1/2	2/3	1/5	1/10	1/5
+ 3/8	+ 1/4	+ 5/8	+ 3/10	+ 2/3	+ 2/9	+ 1/3	+ 2/4	+ 3/4	+ 2/5	+ 2/3

ANSWER: _ _ _ _ _ _ _ _ _ _ _ _ _ _, _ _

NAME________________________

Hi friends,

If laughter is the best medicine, I should be feeling wonderful! While walking down the main street of town, I saw a statue of "Popeye the Sailor." Because so much spinach is grown here, Popeye's their hero, it appears. Do you know where I'm visiting?

Professor Ben Around

To find out, subtract. Write the matching letter above the difference. The letters spell the answer. Some letters will not be used.

ANSWER: $\frac{13}{24}$ __ $\frac{7}{24}$ __ $\frac{13}{18}$ __ $\frac{1}{12}$ __ $\frac{2}{15}$ __ $\frac{5}{12}$ __ $\frac{2}{5}$ __ $\frac{1}{5}$ __ $\frac{7}{16}$ __ $\frac{19}{40}$ __ $\frac{1}{6}$ __, $\frac{11}{36}$ __ $\frac{3}{4}$ __

S	T		
5/12 − 1/3	7/8 − 2/5		

X	P	C
5/6 − 1/12	8/9 − 2/3	2/3 − 1/8

B	I	U
12/15 − 1/5	13/16 − 3/8	3/4 − 3/8

G	Y	O
7/10 − 2/5	2/3 − 1/2	7/9 − 2/7

A	W
3/4 − 1/3	1/3 − 2/27

L	Y	T
3/4 − 7/20	5/6 − 1/9	3/10 − 1/6

H	R	F
2/3 − 1/6	5/8 − 1/3	5/8 − 5/16

T	C
4/9 − 5/36	13/15 − 2/3

NAME______________________________

Hello children,

My niece, Minnie Places, is growing up. On her first date she went to the Iowa State Fair. Her boyfriend won a blue ribbon exhibiting his cow. Do you know what capital city hosts this state fair?

Professor Ben Around

To find out, add or subtract. Write the matching letter above each sum or difference. The letters spell the answer.

ANSWER:

___	___	___		___	___	___	___	___	___	’	___	___
20 1/4	12 1/6	7 2/5		16	4 5/11	4 1/4	2 1/2	10	9 4/9		5 5/12	4 4/7

E
5 1/3 + 4 2/3 =

O
13 8/11 − 9 3/11 =

I
9 8/12 − 4 3/12 =

E
5 3/6 + 6 4/6 =

S
6 17/18 + 2 9/18 =

I
12 3/4 − 8 2/4 =

D
11 10/20 + 8 15/20 =

S
4 4/5 + 2 3/5 =

N
7 7/8 − 5 3/8 =

M
8 8/14 + 7 6/14 =

A
7 5/7 − 3 1/7 =

NAME ______________________

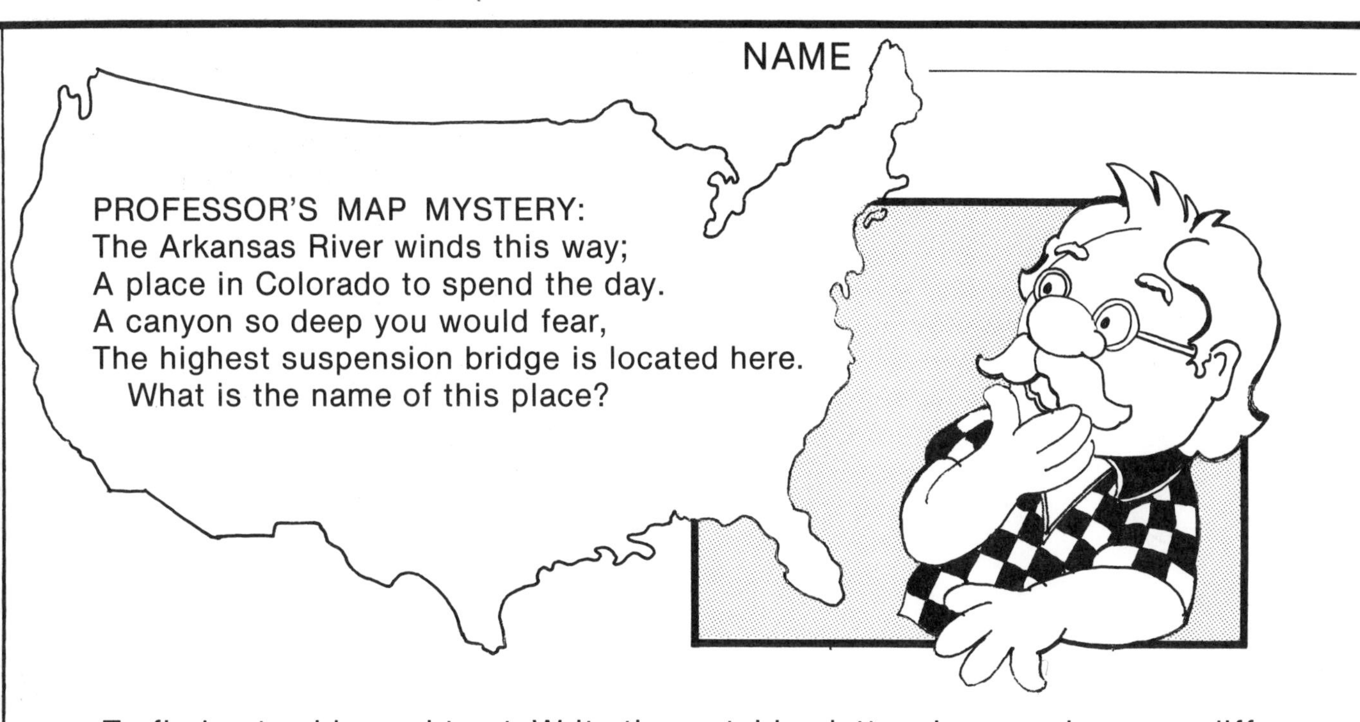

PROFESSOR'S MAP MYSTERY:
The Arkansas River winds this way;
A place in Colorado to spend the day.
A canyon so deep you would fear,
The highest suspension bridge is located here.
What is the name of this place?

To find out, add or subtract. Write the matching letter above each sum or difference. The letters spell the answer.

O	A	G	R	E	O	Y	R	L	G
5 4/5	7 3/4	8 5/10	9 9/18	10 3/6	15 2/3	19 2/10	17 5/7	9 4/15	13 3/8
− 2 3/20	+ 8 5/12	− 3 2/5	+ 8 2/9	− 2 3/24	+ 25 4/6	− 8 3/30	+ 13 2/21	− 5 2/60	+ 7 2/16

ANSWER:

___	___	___	___	___		___	___	___	___	___
17 13/18	41 1/3	11 1/10	16 1/6	4 7/30		5 1/10	3 13/20	30 17/21	20 1/2	8 3/8

NAME____________________

PROFESSOR'S RIDDLE:
Fields of wheat cover my land;
Sunflowers grow in hills of sand.
Vast plains where buffalo roam,
Dwight Eisenhower called me home.
What state am I?

To find out, multiply. Cross out the letter above each product in the decoder. The remaining letters spell the answer. Write the letters in the blanks below.

DECODER

F	I	K	L	D	A	P	C	H
1/24	1/8	3/4	3/16	2/15	9/10	1/3	3/14	1/9
P	N	G	S	O	A	W	H	S
1/4	3/8	1/16	5/9	2/9	2/3	1	5/12	5/8

1. $1/2 \times 1/4 =$ ____
2. $9/12 \times 1/4 =$ ____
3. $6/7 \times 2/8 =$ ____
4. $2/3 \times 3/8 =$ ____
5. $1/9 \times 3/8 =$ ____
6. $6/18 \times 2/3 =$ ____
7. $4/5 \times 1/6 =$ ____
8. $3/8 \times 1/6 =$ ____
9. $4/9 \times 3/4 =$ ____
10. $1/3 \times 3/9 =$ ____
11. $2/3 \times 5/8 =$ ____
12. $5/4 \times 4/5 =$ ____

ANSWER: ____ ____ ____ ____ ____ ____

To find out, multiply. Draw a line from the letter beside each problem to the product. Write the letter in the blank beside the product. The letters spell the answer. Write the letters in the blanks below.

Problem		Letter			Product	Blank
1. 1 2/3 × 2 2/6	=	N	•	•	77 19/28	______
2. 2 3/4 × 3 2/5	=	S	•	•	3 2/9	______
3. 5 2/3 × 2 1/2	=	N	•	•	3 8/9	______
4. 9 1/2 × 2 1/6	=	T	•	•	14 1/6	______
5. 5 3/5 × 2 8/12	=	A	•	•	18 3/8	______
6. 2 5/12 × 1 1/3	=	I	•	•	9 7/20	______
7. 5 1/4 × 3 8/16	=	E	•	•	4 11/16	______
8. 1 1/4 × 3 3/4	=	O	•	•	20 7/12	______
9. 9 3/8 × 8 2/7	=	M	•	•	14 14/15	______

ANSWER: __ __ __ __ __ __ __ __ __

NAME____________________

Professor Ben Around is suffering from jet lag. He decided to get back to nature and spend a few days camping in the mountains of Wyoming. Do you know where he is camping?

To find out, divide. Cross out the letter beside each quotient in the decoder. The remaining letters spell the answer. Write the letters in the blanks below.

1. 1/2 ÷ 1/3 = ____

2. 2/6 ÷ 5/8 = ____

3. 4/5 ÷ 3/9 = ____

4. 7/8 ÷ 2/3 = ____

5. 6/7 ÷ 1/8 = ____

6. 3/5 ÷ 2/6 = ____

7. 3/8 ÷ 3/4 = ____

8. 3/4 ÷ 1/8 = ____

9. 3/7 ÷ 4/5 = ____

10. 5/6 ÷ 1 = ____

11. 2/5 ÷ 2/3 = ____

12. 1/12 ÷ 1/8 = ____

13. 7/10 ÷ 5/8 = ____

14. 1/4 ÷ 3/5 = ____

15. 1/2 ÷ 3/20 = ____

16. 4/5 ÷ 2/5 = ____

17. 2/8 ÷ 4/9 = ____

18. 2/3 ÷ 1/6 = ____

DECODER

F	2 2/5
L	9/16
G	5/8
S	1/2
O	1 3/25
R	3/4
H	3/5
T	1 5/16
A	1/5
L	1 1/2
N	2 3/8
I	6
Y	5/12
D	7/8
C	2
J	8/15
T	7/11
W	2/3
E	7/12
B	15/28
T	1/3
M	6 6/7
O	1/8
P	5/6
N	9/10
K	3 1/3
S	3
U	1 4/5
H	4

ANSWER: __ __ __ __ __   __ __ __ __ __ __

NAME______________________________

To find out, divide. Write the matching letter in the blank above the quotient. The letters spell the answer.

ANSWER:

___	___	___	___	___	___	' ___		___	___	___	___	___	___	___	___
2 4/63	13/22	2 17/20	3 4/5	3/8	27/56	19/46		21/25	2 16/25	31/40	1 5/14	45/92	4/5	2 17/64	3 15/32

3 4/5 ÷ 1 1/3	R =	______	2 3/8 ÷ 1 3/4	E =	______
4 3/4 ÷ 1 1/4	T =	______	4 1/5 ÷ 5	V =	______
8 2/3 ÷ 4 1/5	M =	______	6 3/5 ÷ 2 1/2	I =	______
3 1/4 ÷ 5 1/2	A =	______	2 7/12 ÷ 3 1/3	N =	______
4 5/8 ÷ 1 3/9	D =	______	1 7/8 ÷ 3 5/6	Y =	______
1 1/2 ÷ 4	H =	______	2 3/8 ÷ 5 6/8	S =	______
1 1/8 ÷ 2 2/6	A =	______	3 5/8 ÷ 1 3/5	R =	______
1 1/5 ÷ 1 1/2	A =	______			

PROFESSOR'S RIDDLE:
The father of our country
Was a surveyor here,
But today my famous apples
Are what I hold so dear.
What city am I?

NAME ____________________

To find out, match the words to the standard numerals on the apples. Write the matching letters in the blanks above the numbers of the problems. The letters spell the answer.

ANSWER: 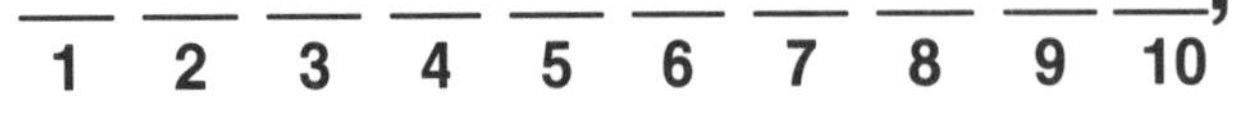

__ __ __ __ __ __ __ __ __ __,
1 2 3 4 5 6 7 8 9 10

__ __ __ __ __ __ __ __
11 12 13 14 15 16 17 18

1. Three and seven tenths
2. One and eight tenths
3. Four and six hundredths
4. Two and two hundredths
5. Four and six thousandths
6. Two and two thousandths
7. Two and two tenths
8. Five and nine hundredths
9. Three and seven thousandths
10. Six and four thousandths
11. Four and six tenths
12. One and eight thousandths
13. Six and four hundredths
14. Six and four tenths
15. Five and nine tenths
16. Five and nine thousandths
17. One and eight hundredths
18. Three and seven hundredths

NAME________________________

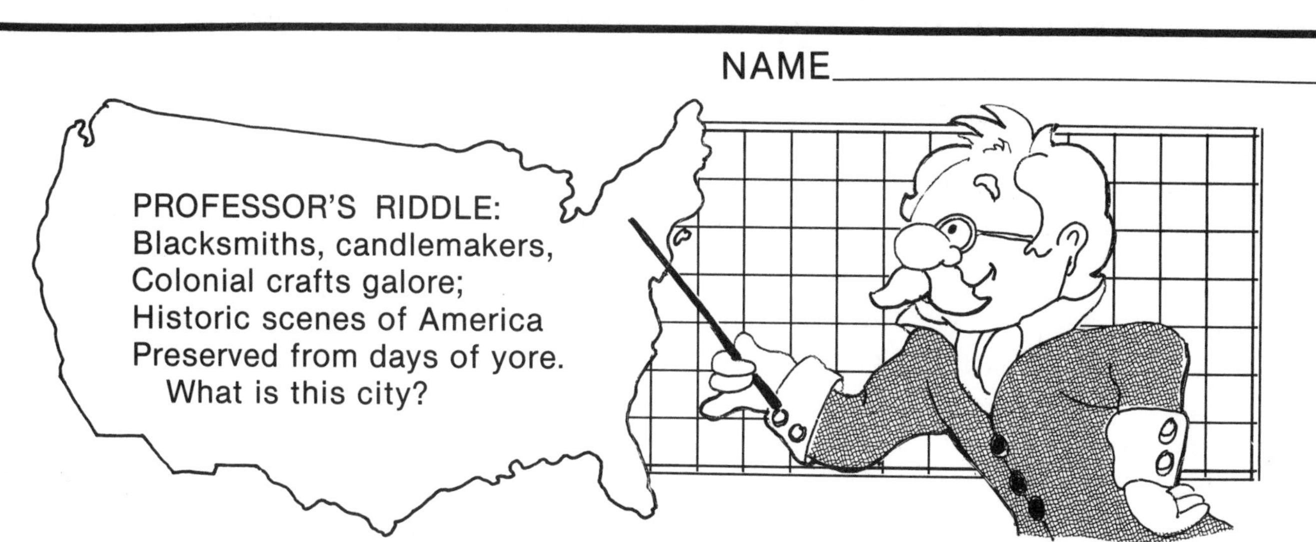

To find out, circle the letter above the largest decimal in each group. The circled letters spell the answer. Write the letters in the blanks below.

1.	S 3.05	W 3.50	C 3.005
2.	H 1.03	O 1.035	I 1.3
3.	L .46	D .406	A .046
4.	I 2.9	R 2.09	L 2.99
5.	I .33	H .303	L .033
6.	N 5.8	E 5.08	A 5.808
7.	M 1.20	S 1.02	A 1.002
8.	S 3.76	T 3.076	P 3.706
9.	D 7.20	O 7.025	B 7.25
10.	U .47	N .407	L .047
11.	R .25	A .025	S .205
12.	C .601	D .061	G .610
13.	I .512	V 1.25	T .521
14.	I 1.40	N 1.04	P .401
15.	S 2.09	R 2.90	I .920
16.	G .350	H .053	L .035
17.	I 1.07	S .071	E .710
18.	R 3.36	N 3.63	S .633
19.	T .187	I 1.78	W .871
20.	A 6.50	H 6.05	R .650

ANSWER: __ __ __ __ __ __ __ __ __ __ __ __,

__ __ __ __ __ __ __ __

PROFESSOR'S RIDDLE:
Daring acts on the high trapeze,
Animals, clowns, and jugglers to please.
The Big Top Circus of the Ringling Brothers
Began in this town to entertain others.
What is this city?

NAME________________________________

To find out, circle the letter above the decimal that comes between each pair of numerals. Write the circled letter in the blank above the number of the problem. The letters spell the answer.

ANSWER: __ __ __ __ __ __ __, __ __ __ __ __ __ __ __ __
1 2 3 4 5 6 7 8 9 10 11 12 13 14 15 16

1. 5 and 6
B C D
5.2 .56 6.0

2. 3 and 4
O I A
3.0 .34 3.5

3. 1 and 3
E R L
1.0 2.5 3.1

4. 6 and 7
A L O
6.2 .67 7.0

5. 5 and 8
M A B
.58 8.2 5.4

6. 8 and 10
O S B
8.7 10.1 10.8

7. 9 and 12
U O T
12.9 11.5 12.0

8. 7 and 8
S E W
.78 .87 7.8

9. 4 and 6
I X O
5.0 6.0 4.0

10. 2 and 4
H A S
.24 2.0 2.4

11. 3 and 5
I C S
.53 3.5 .35

12. 9 and 10
S O A
9.0 9.1 10.0

13. 5 and 7
N A B
5.6 7.0 5.0

14. 6 and 9
S C E
7.5 9.0 9.6

15. 1 and 2
A I R
1.0 1.5 2.1

16. 8 and 9
N T W
8.9 9.8 9.0

NAME______________________________

Granny Gone Around is spending a few days in a health resort city in the Mountain State. Its spring waters have been used for bathing since George Washington first surveyed the land for Lord Fairfax in the late 1700's. Where is Granny visiting?

To find out, round each decimal to the nearest whole number. Draw a line from the letter beside each decimal to its rounded answer. Write the letter in the blank beside the rounded numeral. The letters spell the answer. Write the letters in the blanks below.

1.	5.47	E	•	• 17 _____
2.	3.28	P	•	• 5 _____
3.	10.6	V	•	• 16 _____
4.	4.38	G	•	• 9 _____
5.	12.8	R	•	• 18 _____
6.	9.56	S	•	• 19 _____
7.	8.3	N	•	• 14 _____
8.	6.87	W	•	• 12 _____
9.	6.25	S	•	• 6 _____
10.	16.7	B	•	• 3 _____
11.	14.4	E	•	• 13 _____
12.	8.96	K	•	• 15 _____
13.	11.5	Y	•	• 8 _____
14.	14.58	I	•	• 4 _____
15.	15.5	R	•	• 10 _____
16.	17.8	E	•	• 7 _____
17.	19.3	L	•	• 11 _____

ANSWER:

_ _ _ _ _ _ _ _ _ _ _ _ _ _ _ _ _ _ _, _ _

NAME ____________________

SPECIAL BULLETIN!

Baby Go Too has wandered away from Professor Ben Around's campsite in the national park that is home to the world's largest herd of Roosevelt elk. Your help is needed to assist the rescue team in finding Baby Go Too.

By working the problems correctly, you can find Baby Go Too and discover the name of the campsite.

Begin at the campsite and follow the path to Baby Go Too by adding .3 to each answer. Write the letters from the path in the blanks below to spell the name of the recreational area.

ANSWER: _, _ _

CAMPSITE

1.2	O 1.5	L 1.8	Y 2.0	E 2.3	L 2.5	S 3.6	M 3.9	B 4.1
S 1.7	N 2.2	Y 2.1	L 2.3	I 2.6	N 2.9	A 3.2	T 4.3	L 4.6
E 2.0	Q 2.3	M 2.4	P 2.7	T 3.1	S 3.4	R 3.7	K 4.0	A 4.9
O 2.9	U 2.6	C 3.3	I 3.0	J 4.0	O 4.8	N 5.1	A 5.4	C 5.1
E 5.8	R 5.5	N 3.6	A 3.9	T 4.2	I 4.5	B 5.0	L 5.7	T 5.4
D 6.1	K 5.2	R 4.9	E 4.0	W 5.9	B 5.6	N 5.3	P 6.0	E 6.1
W 6.4	O 6.7	A 4.6	C 4.3	R 6.1	T 6.4	R 6.6	A 6.3	I 6.5
D 6.2	O 7.0	P 4.2	A 4.5	C 6.2	M 6.7	K 6.9	W 7.2	A 7.5

BABY GO TOO

NAME___________________________

While visiting the Prairie State, Professor Ben Around enjoyed a day at the zoo. His favorite place was "Baboon Island" where baboons romp and play. Where is the famous zoo he was visiting?

To find out, subtract. Cross out the letters above each difference in the decoder. The remaining letters spell the answer. Write the letters in the blanks below.

DECODER

S 22.6	A 18.9	N 26.2	B 14.9	D 12.2	R 16.5	I 27.3	E 14.5
G 13.1	P 20.5	O 18.1	O 35.1	R 21.1	T 19.9	K 22.1	F 13.6
L 30.8	I 30.3	L 22.7	Y 16.8	E 17.0	A 25.4	L 37.9	N 17.8
N 25.8	M 32.5	D 20.9	A 21.8	I 23.7	C 31.7	L 42.8	Y 21.9

ANSWER: __ __ __ __ __ __ __ __ __ __, __ __

1. $29.5 - 8.4$

2. $36.8 - 9.5$

3. $24.3 - 6.5$

4. $30.4 - 8.5$

5. $28.3 - 5.7$

6. $31.4 - 8.7$

7. $18.9 - 6.7$

8. $23.4 - 8.9$

9. $35.3 - 9.5$

10. $22.5 - 5.7$

11. $29.8 - 9.9$

12. $38.2 - 6.5$

13. $21.6 - 8.5$

14. $27.9 - 9.0$

15. $26.3 - 4.5$

16. $32.8 - 7.4$

17. $35.6 - 9.4$

18. $41.0 - 8.5$

19. $38.6 - 7.8$

20. $29.1 - 8.6$

PROFESSOR'S MAP MYSTERY:
Waterfalls are fascinating!
Here's a clue for all:
In the state of Kentucky,
It's the highest waterfall.
What is its name?

NAME ______________________

To find out, add. Circle the letter above the matching sum in the decoder. The circled letters from left to right spell the answer. Write the letters in the blanks below.

1. 12.5 + 3.8

2. 21.3 + 4.5

3. 18.6 + 5.8

4. 14.7 + 5.2

5. 26.35 + 7.43

6. 34.56 + 9.23

7. 19.58 + 5.74

8. 36.03 + 4.29

9. 40.18 + 9.38

10. 39.24 + 5.38

11. 28.06 + 8.74

12. 15.43 + 5.29

13. 15.46 + 32.07 + 12.29

14. 26.43 + 10.39 + 19.44

15. 30.28 + 11.43 + 12.97

ANSWER: ___ ___ ___ ___ ___ ___ ___ ___ ___ ___

___ ___ ___ ___ ___

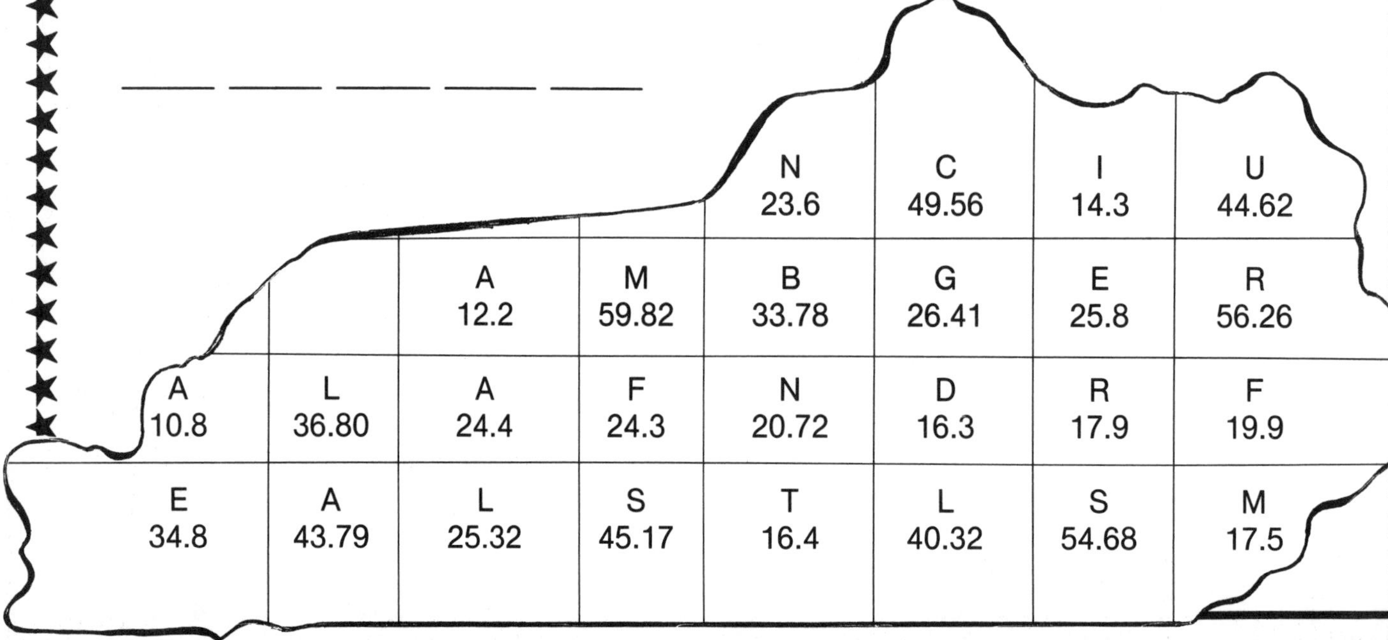

PROFESSOR'S MAP MYSTERY: NAME________________________

Sailing! Sailing!
South of the Hoosier State
On the state's southern border,
Come aboard and be my shipmate!

What is my location?
Professor Ben Around

To find out, subtract. Cross out the letter above each difference inside the state outline. The remaining letters spell the answer. Write the letters in the blanks below.

1. $217.5 - 38.93$

2. $386.53 - 29.24$

3. $543.8 - 39.346$

4. $426.2 - 93.7$

5. $302.56 - 49.3$

6. $817.3 - 29.46$

7. $305.217 - 39.348$

8. $246.173 - 93.805$

9. $458.251 - 96.482$

10. $329.251 - 48.14$

11. $132.7 - 89.173$

ANSWER:

__ __ __ __

__ __ __ __ __

DECODER

T 504.454	O 205.31	E 361.769
N 152.368	H 638.57	N 178.57
I 186.94	E 357.29	O 302.19
S 253.26	R 706.81	E 332.5
I 514.52	B 787.84	V 308.54
E 243.16	R 38.910	A 265.869
O 281.111	S 43.527	

NAME______________________

Anchors away!

Here I am at the largest naval base in the United States. Seeing Navy ships on Chesapeake Bay brings back glorious memories of my military days. What naval base am I visiting?

Professor Ben Around

U.S. POSTAGE 0 1 2 5

To find out, multiply. Cross out the letter beside each anchor that contains a product. The remaining letters spell the answer. Write the letters in the blanks below.

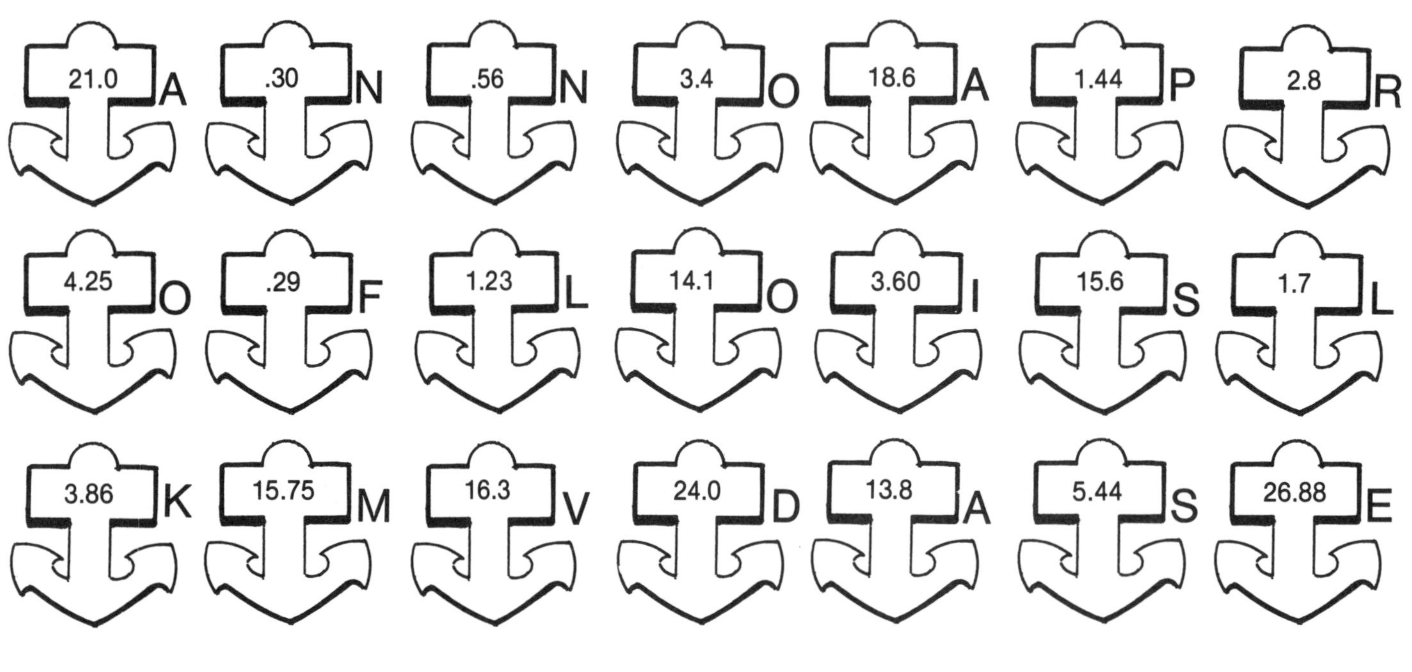

1. 3.5×6
2. 4.8×5
3. 6.2×3
4. 7.8×2
5. $1.5 \times .2$
6. $3.6 \times .4$
7. $7.2 \times .5$
8. $4.1 \times .3$
9. 1.7×2.5
10. 3.4×1.6
11. 6.3×2.5
12. 8.4×3.2

ANSWER: ___ ___ ___ ___ ___ ___ ___, ___ ___

NAME________________________

PROFESSOR'S RIDDLE:
The world's longest bridge
Spans this lake.
It's a beautiful body of water
For a recreational break.
What is the name of the lake?

To find out, multiply. Write the matching letter in the blank above the product. The letters spell the answer.

ANSWER:

___	___	___	___
10,643.58	3,207.96	720.174	59.5

___	___	___	___	___	___	___	___	___	___	___	___	___
1,466.994	1,019.35	642.642	4,155.36	15,516.86	1,151.018	290.852	6,968.5	55.971	752.4	203.2044	1,166.2	59.3217

1. 17.5 × 3.4 = E ________
2. 207.3 × .27 = T ________
3. 49.5 × 15.2 = R ________
4. 39.57 × 18.2 = K ________
5. 416.5 × 2.8 = I ________
6. 509.2 × 6.3 = A ________
7. 611.7 × 17.4 = L ________
8. 50.93 × 22.6 = H ________
9. 107.3 × 9.5 = O ________
10. 23.54 × 27.3 = N ________
11. 76.54 × 3.8 = A ________
12. 157.4 × 26.4 = T ________
13. 398.2 × 17.5 = R ________
14. 520.7 × 29.8 = C ________
15. 34.29 × 1.73 = N ________
16. 56.92 × 3.57 = A ________
17. 84.31 × 17.4 = P ________

NAME ____________________

PROFESSOR'S RIDDLE:
Cotton is this state's king.
Magnolia trees bloom in spring.
Southern belles grace this place;
Here one finds a famous trace.
What is this state?

To find out, multiply. Cross out the letter beside each product in the decoder. The remaining letters spell the answer. Write the letters in the blanks below.

1. 25.34 × .82	2. 17.36 × .32	3. 24.58 × .54	4. 26.31 × .35
5. 581.4 × .36	6. 321.4 × .52	7. 876.3 × .43	8. 953.2 × .37
9. 257.3 × 14.2	10. 348.9 × 22.3	11. 513.2 × 26.5	12. 186.3 × 19.4
13. 581.7 × 12.3	14. 861.3 × 11.4	15. 638.4 × 13.5	16. 702.8 × 22.6
17. 578.36 × 3.2	18. 802.49 × 5.6	19. 392.17 × 4.8	20. 157.28 × 5.4

DECODER

A	352.684
F	13.2732
M	25.461
R	1,850.752
I	3,643.21
G	20.7788
E	4,493.944
L	376.809
S	142.78
K	7,780.47
S	3,437.31
O	13,599.8
I	204.73
A	15,883.28
N	5.5552
S	438.64
R	3,614.22
S	167.128
G	1,882.416
O	9.2085
S	8.3176
A	7,154.91
I	5,431.72
P	311.464
S	209.304
R	8,618.4
P	7,293.8
I	849.312
I	502.170
D	3,653.66
A	9,818.82

ANSWER: _ _ _ _ _ _ _ _ _ _ _ _ _ _ _ _ _ _

NAME ______________________

PROFESSOR'S MAP MYSTERY:
The United Nations makes its home here.
World famous headquarters for peace and good cheer.
What is this city?

To find out, divide. Shade each space that contains a quotient less than the whole number one. The shaded area outlines the state where the U.N. is located. Unscramble the letters in the remaining spaces to spell the name of the city. Write the letters in the blanks.

ANSWER: ___ ___ ___   ___ ___ ___ ___   ___ ___ ___ ___

★★★★★★★★★★★★★★★★★★★★★★★★★★★★★★★★★★★★★

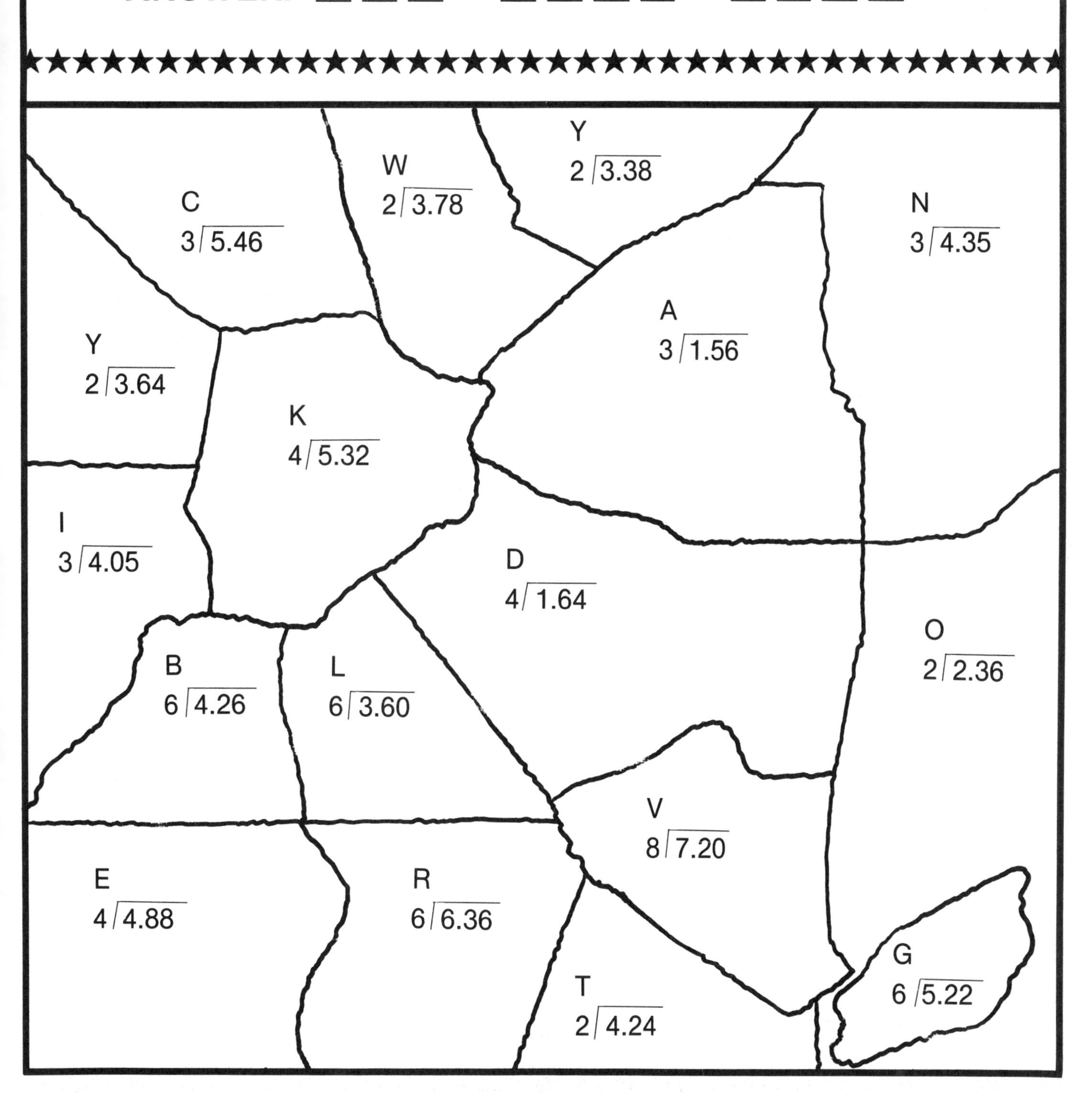

To find out, divide. Arrange the quotients in order from least to greatest. The letters of the ordered quotients spell the answer. Write the letters in the blanks below.

D $.5\overline{)7.4}$	A $.3\overline{)6.12}$	N $.4\overline{)8.72}$
N $.6\overline{)7.35}$	I. $.4\overline{).08}$	I $.6\overline{)9.24}$
L $.26\overline{)32.5}$	P $.39\overline{)40.17}$	A $.45\overline{)45.9}$
I $.25\overline{)32.5}$	O $.17\overline{)19.04}$	N $.16\overline{)82.4}$
S $.32\overline{)44.80}$	I $.24\overline{)42.6}$	

ANSWER:

____ ____ ____ ____ ____ ____ ____ ____ ____ ____ ____ ____,

____ ____

NAME________________________

Hello Boys and Girls,

Everything is coming up roses in the largest city of this Pacific Coast state. The city has been given the nickname "City of Roses" because of its beautiful rose gardens. What is its name?

Professor Ben Around

To find out, divide. Cross out the letter beside each quotient in the decoder. The remaining letters spell the answer. Write the letters in the blanks below.

1. $56\overline{)201.6}$
2. $43\overline{)270.9}$
3. $97\overline{)407.4}$
4. $32\overline{)156.8}$
5. $25\overline{)232.5}$
6. $71\overline{)369.2}$
7. $6.5\overline{)20.15}$
8. $4.2\overline{)15.96}$
9. $7.3\overline{)28.47}$
10. $2.3\overline{)13.11}$
11. $5.8\overline{)42.34}$
12. $3.5\overline{)23.45}$

DECODER

P	4.6
S	3.9
A	3.8
N	6.3
O	5.8
F	9.3
R	2.7
A	4.9
T	2.5
N	3.6
C	6.7
L	9.6
I	7.3
A	3.3
S	4.2
N	7.5
D	6.0
C	3.1
O	4.5
C	5.7
A	5.2
R	4.3

ANSWER: ___________________________, ________

NAME____________________________

PROFESSOR'S RIDDLE:
Moose and timber wolves
Make their homes here.
On Michigan's largest island,
They've nothing to fear.
What is the name of the island?

To find out, solve each problem. Cross out the letter above each answer in the decoder. The remaining letters spell the name of the island. Write the letters in the blanks below.

1. $17.4 + 38.5$

2. $125.7 - 38.9$

3. $469.5 + 231.6$

4. $43.5 - 9.64$

5. 4.5×0.3

6. 3.8×5.6

7. 27.5×1.3

8. 246.27×1.3

9. $5\overline{)4.55}$

10. $.3\overline{)1.896}$

11. $4.3\overline{)28.81}$

12. $.25\overline{)26.25}$

DECODER

M	E	I	C	R	S	H	T	L	I	E
.91	320.151	.84	55.9	105	62.5	701.1	86.8	38.4	21.28	37.6
R	G	O	A	Y	N	A	D	A	L	E
.78	6.32	5.14	35.75	42.18	6.7	8.3	1.35	33.86	14.6	15.9

ANSWER: ____ ____ ____ ____

____ ____ ____ ____ ____ ____

PROFESSOR'S GEOGRAPHIC SKILLS

As Professor Ben Around flips through his photograph albums, he is filled with love for his homeland, the United States of America. He is impressed by the diversity of landscapes, cities, national parks, climate, and natural life. Truly, there is no place like home! Help Professor Ben Around remember his favorite places by completing the activities in the next section.

NAME___________________________

TRACKING THE PROFESSOR

An avid traveler, Professor Ben Around is ready to depart on a moment's notice. Use the map of the United States on page 93 to track Professor Ben Around as he explores the scenic USA. Find each starting point, follow the trail, and write the professor's destination in the blank.

1. STARTING POINT: North Carolina
   TRAIL: 1 state west, 1 state north
   DESTINATION: ____________________
2. STARTING POINT: Texas
   TRAIL: 2 states north, 1 east, 1 south
   DESTINATION: ____________________
3. STARTING POINT: Kansas
   TRAIL: 1 state east, 2 south, 1 west
   DESTINATION: ____________________
4. STARTING POINT: Wisconsin
   TRAIL: 1 state south, 1 east, 2 south
   DESTINATION: ____________________
5. STARTING POINT: Oklahoma
   TRAIL: 1 state north, 2 west, 1 south
   DESTINATION: ____________________
6. STARTING POINT: Vermont
   TRAIL: 1 state east, 1 south, 1 west
   DESTINATION: ____________________
7. STARTING POINT: Tennessee
   TRAIL: 1 state east, 1 south, 2 west
   DESTINATION: ____________________
8. STARTING POINT: Colorado
   TRAIL: 1 west, 1 south, 1 east
   DESTINATION: ____________________
9. STARTING POINT: Missouri
   TRAIL: 2 north, 1 east, 1 south, 1 east
   DESTINATION: ____________________
10. STARTING POINT: Arizona
    TRAIL: 1 north, 2 west, 1 north, 1 east
    DESTINATION: ____________________

The Professor is in Washington, D.C. Can you give him directions to your state? Fill in the trail and destination to help him visit your area.

STARTING POINT: Washington, D.C.

TRAIL: ____________________________

DESTINATION: ______________________

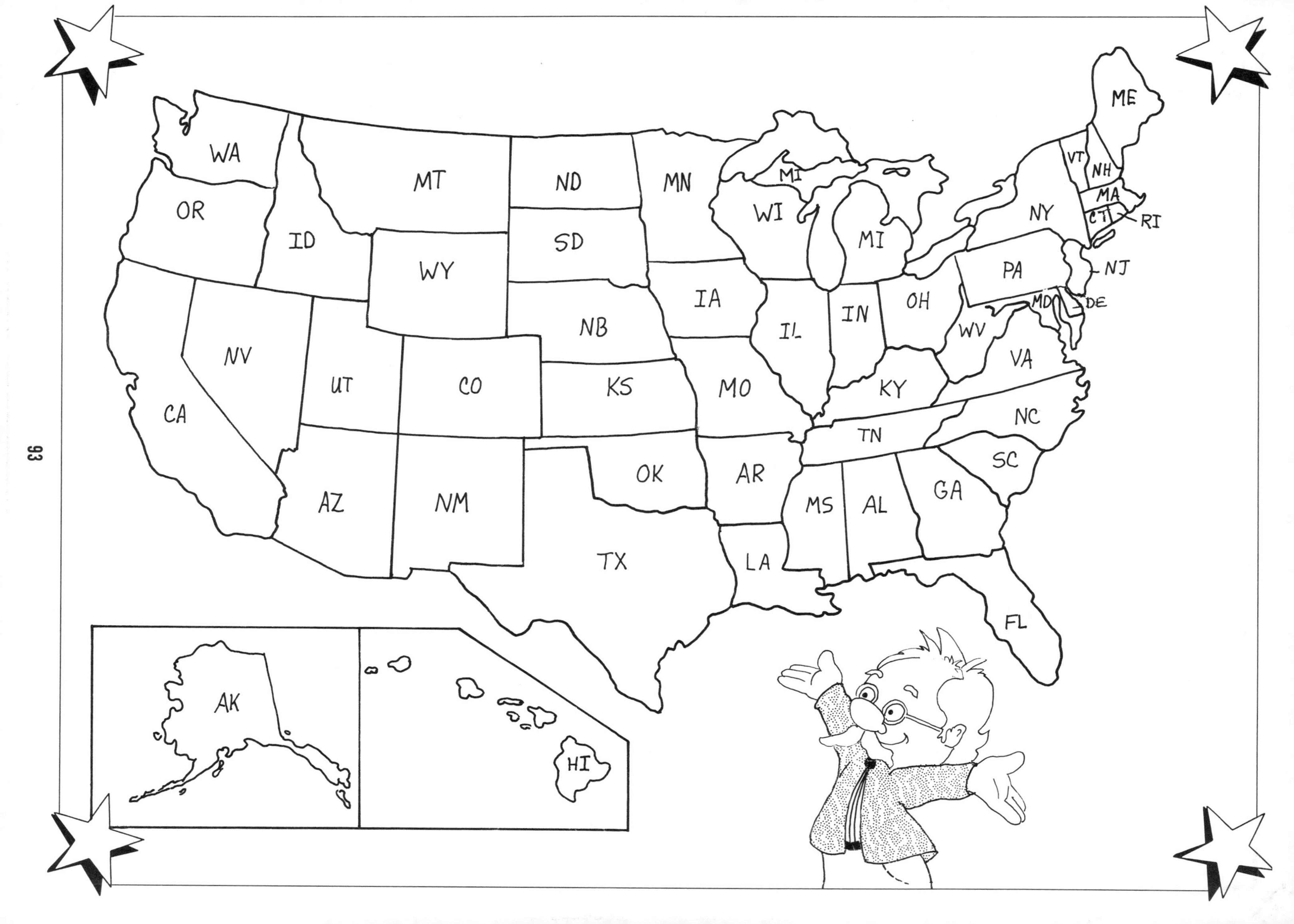
WA
OR
ID
MT
ND
MN
MI
WI
MI
NY
VT
NH
ME
MA
CT
RI
SD
WY
PA
NJ
MD
DE
IA
NB
IL
IN
OH
WV
NV
UT
CO
KS
MO
KY
VA
CA
NC
TN
OK
AR
SC
AZ
NM
MS
AL
GA
TX
LA
FL
AK
HI

NAME ____________________

MAP OF STATES AND CAPITALS OF THE USA

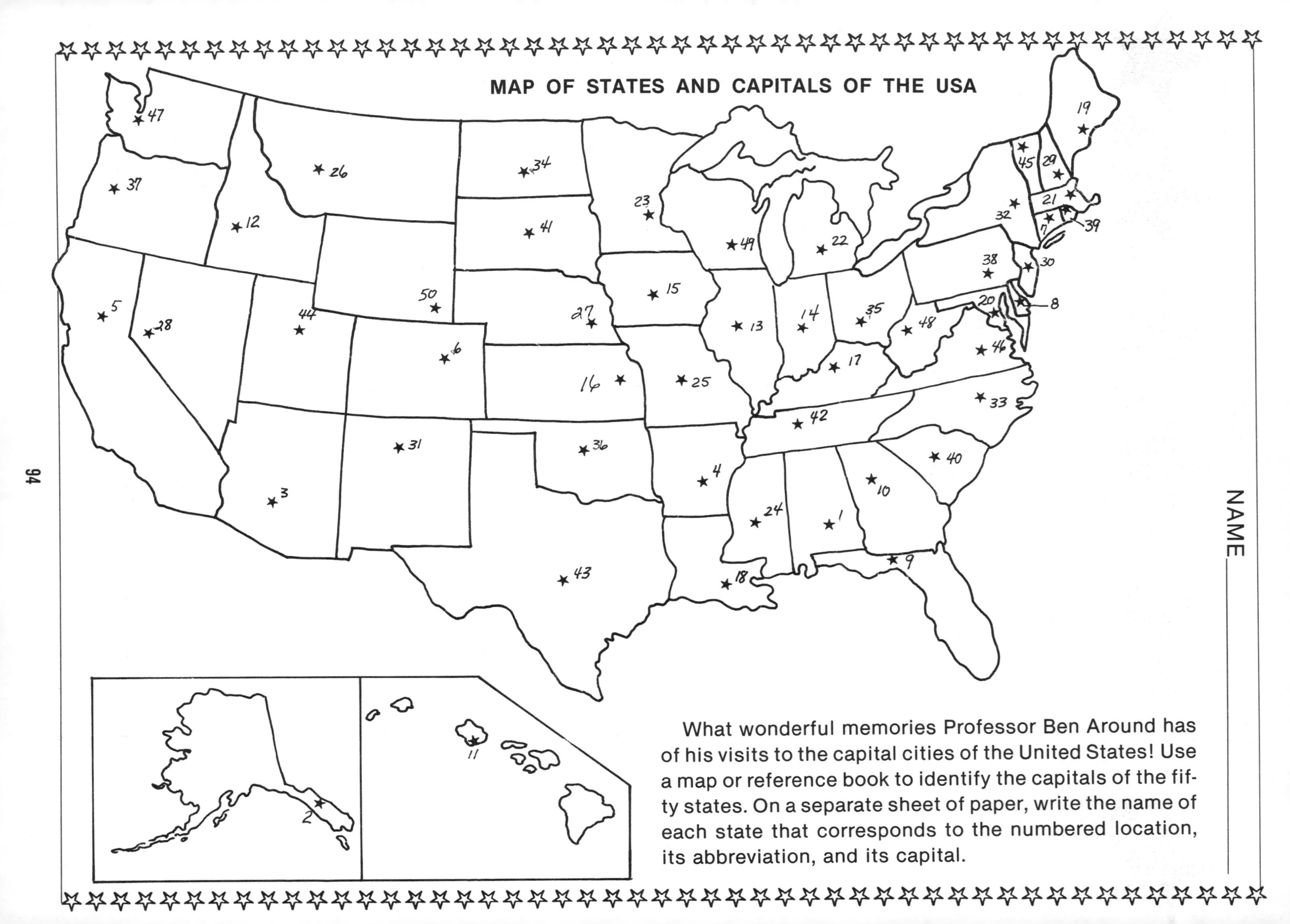

What wonderful memories Professor Ben Around has of his visits to the capital cities of the United States! Use a map or reference book to identify the capitals of the fifty states. On a separate sheet of paper, write the name of each state that corresponds to the numbered location, its abbreviation, and its capital.

GEOGRAPHICAL TERMS

While touring from state to state, Professor Ben Around discovered many regional differences. The variety of landscapes brought a flood of geographical terms to mind. The puzzle below contains 25 geographical words that should be familiar to you. How many can you find? You may look horizontally, vertically, and diagonally. Circle your answers.

S C H A N T L O C B E T R U E N A V S L D

F I J L A N D F O R M J K S L F M O P O T

C R T W E Y S X R I V E R E Z B S Y U N L

H E Z I S T H M U S B Z J A E Z S O R G A

Y S T P L A T E A U L G N D U K T I Y I K

F E R Z S B H I Y M A Q U B H L R E D T S

A R T S R E N A J F K T B L A N A S E U W

J V B D M W B S K X I W S Y F W I N L D R

B O T V A L L E Y T U S I T A Z T R T E O

E I R S Y T Q U A C A N Y O N F J O A B S

X R I A D J K L T X W S F P L A I N W Y P

H E B Y W G B A S I N H Z J B O B R P S E

T S U A V M O U N T A I N A D G A Z R A N

A Z T Q U A N X F R Q V S N W S D B A E I

C H A G I N L E T H S Y A K I O J L I Y N

I J R N K R W E I O T L T Z S U E A R O S

B S Y E A S N O X F S G D A Y R G K I S U

M E R I D I A N W I Z H Q U K C O E E F L

H O N S K V E Z L M Y R N P O E Y N X A A

WORD BANK

bay	canyon	tributary	gulf	basin
sea	island	plateau	plain	river
valley	inlet	source	strait	latitude
reservoir	peninsula	longitude	lake	mountain
meridian	isthmus	delta	prairie	landform

NAME________________________________

GEOGRAPHICAL TERMS

Show Professor Ben Around you know your geographical terms by placing the letter of the correct definition in the blank beside each word.

____	1. basin	A. the place where a river begins
____	2. bay	B. land surrounded by water
____	3. canyon	C. broad, flat land
____	4. delta	D. a region drained by a river and its tributaries
____	5. gulf	E. lowland between hills or mountains
____	6. island	F. part of a body of water reaching into the land; usually smaller than a gulf
____	7. isthmus	G. land almost surrounded by water
____	8. lake	H. flat land that is raised higher than the surrounding land
____	9. mountain	I. a deep valley with steep sides
____	10. mouth	J. land formed at the mouth of a river by deposited silt, sand, and pebbles
____	11. peninsula	K. a river or stream that flows into a larger body of water
____	12. plain	L. a body of water larger than a bay that reaches into the land
____	13. plateau	M. an inland body of water
____	14. reservoir	N. a narrow waterway connecting two larger bodies of water
____	15. source	O. a narrow piece of land bordered by water and connecting two larger pieces of land
____	16. strait	P. high, rocky land that is higher than a hill
____	17. tributary	Q. the place where a river flows into a larger body of water
____	18. valley	R. a natural or man-made place for storing water

NAME________________________

BEN'S BINGO

Professor Ben Around plays Bingo in an exciting way! Join in the fun by writing twenty-four states in the spaces on your Bingo sheet. Your teacher will call out an abbreviation of a state. Cover up the state if it is on your card. Continue playing until someone covers five spaces in a horizontal, vertical, or diagonal row and yells "Bingo!"

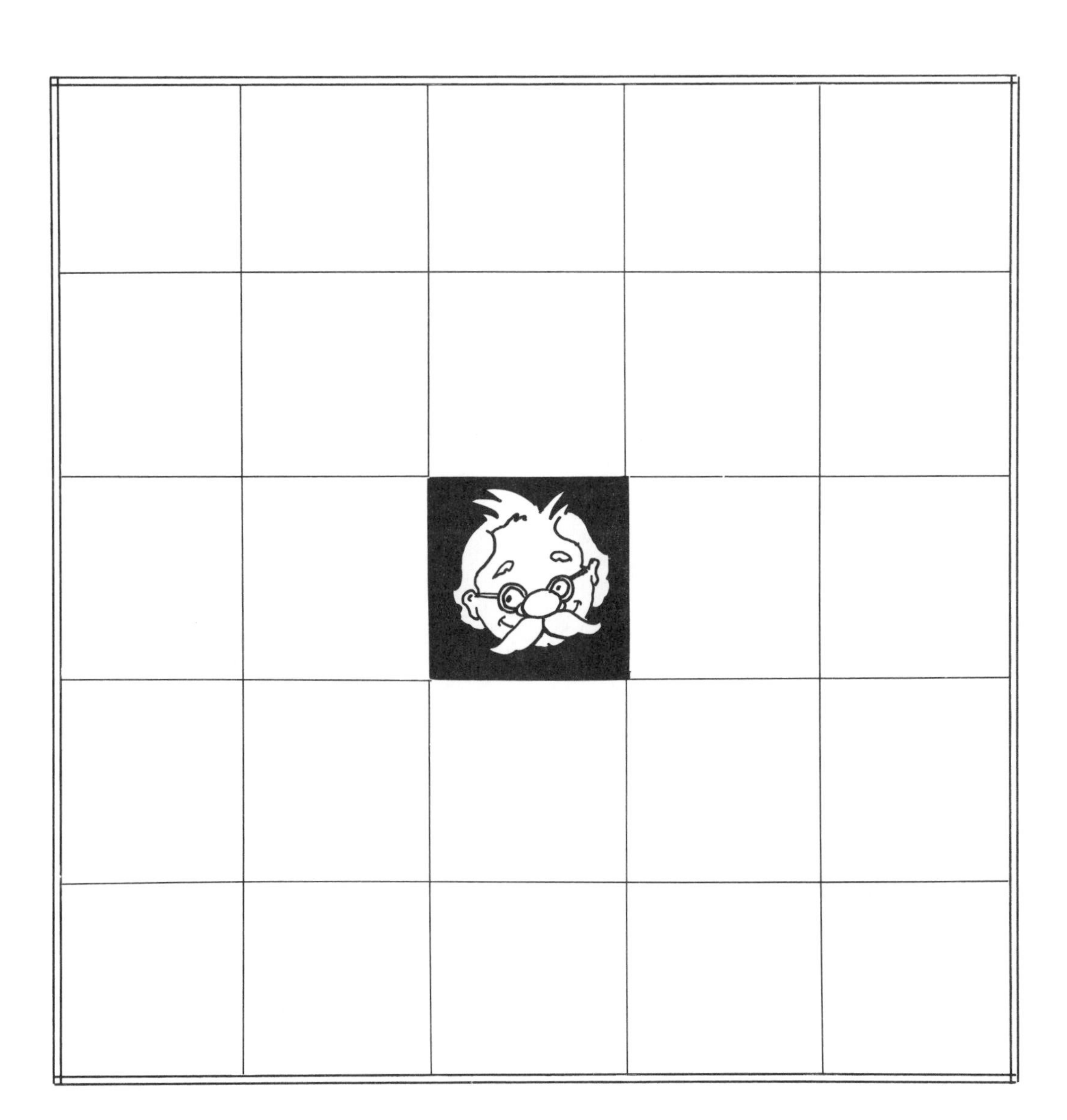

Variation: Write state abbreviations on your Bingo sheet, and the teacher will call out the states.

STATES AND ABBREVIATIONS

ALABAMA	AL	MONTANA	MT
ALASKA	AK	NEBRASKA	NE
ARIZONA	AZ	NEVADA	NV
ARKANSAS	AR	NEW HAMPSHIRE	NH
CALIFORNIA	CA	NEW JERSEY	NJ
COLORADO	CO	NEW MEXICO	NM
CONNECTICUT	CT	NEW YORK	NY
DELAWARE	DE	NORTH CAROLINA	NC
FLORIDA	FL	NORTH DAKOTA	ND
GEORGIA	GA	OHIO	OH
HAWAII	HI	OKLAHOMA	OK
IDAHO	ID	OREGON	OR
ILLINOIS	IL	PENNSYLVANIA	PA
INDIANA	IN	RHODE ISLAND	RI
IOWA	IA	SOUTH CAROLINA	SC
KANSAS	KS	SOUTH DAKOTA	SD
KENTUCKY	KY	TENNESSEE	TN
LOUISIANA	LA	TEXAS	TX
MAINE	ME	UTAH	UT
MARYLAND	MD	VERMONT	VT
MASSACHUSETTS	MA	VIRGINIA	VA
MICHIGAN	MI	WASHINGTON	WA
MINNESOTA	MN	WEST VIRGINIA	WV
MISSISSIPPI	MS	WISCONSIN	WI
MISSOURI	MO	WYOMING	WY

NAME________________________________

Professor Ben Around was cleaning out his attic. Inside a dusty, old trunk he found what appeared to be a treasure map. Use a ruler to help the professor calculate the number of miles to the treasure.

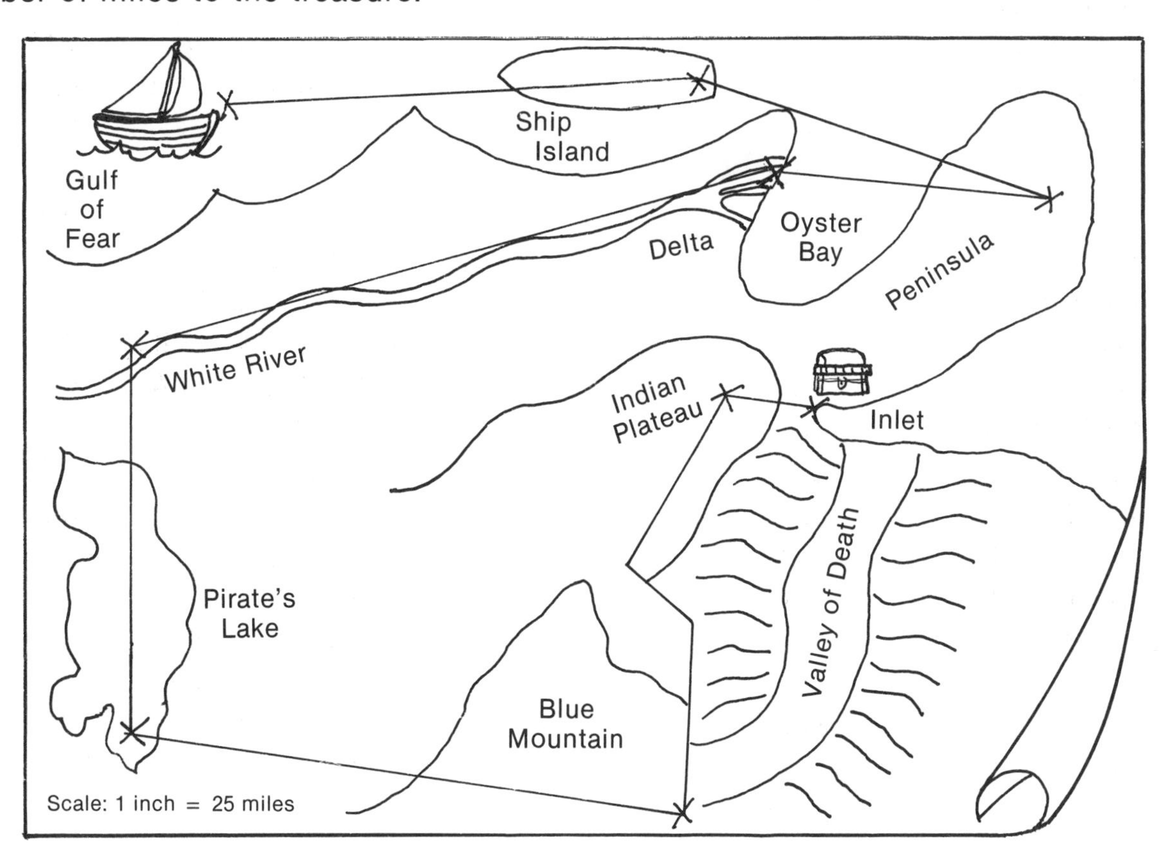

Professor Ben will have to travel _______ miles from the Gulf of Fear to Ship Island. He then must go _______ miles across Oyster Bay to the peninsula. Once again he must cross Oyster Bay _______ miles to the delta. From there he goes _______ miles up the White River. From the river he goes _______ miles to Pirate's Lake. His next destination is _______ miles to the base of Blue Mountain. Professor Ben must avoid the Valley of Death and travel _______ miles to the top of Indian Plateau. His treasure will be found _______ miles away at the inlet. What is the total distance in miles that the professor will travel to find the treasure? _______ miles

If Professor Ben thinks he can travel 40 miles a day, how many miles will it take him to reach the treasure? (Round off the answer to the nearest day.) _______ days

NAME________________________________

Determine the latitude and longitude of each point. Connect the points in order of the problems to discover the island where Professor Ben will retire.

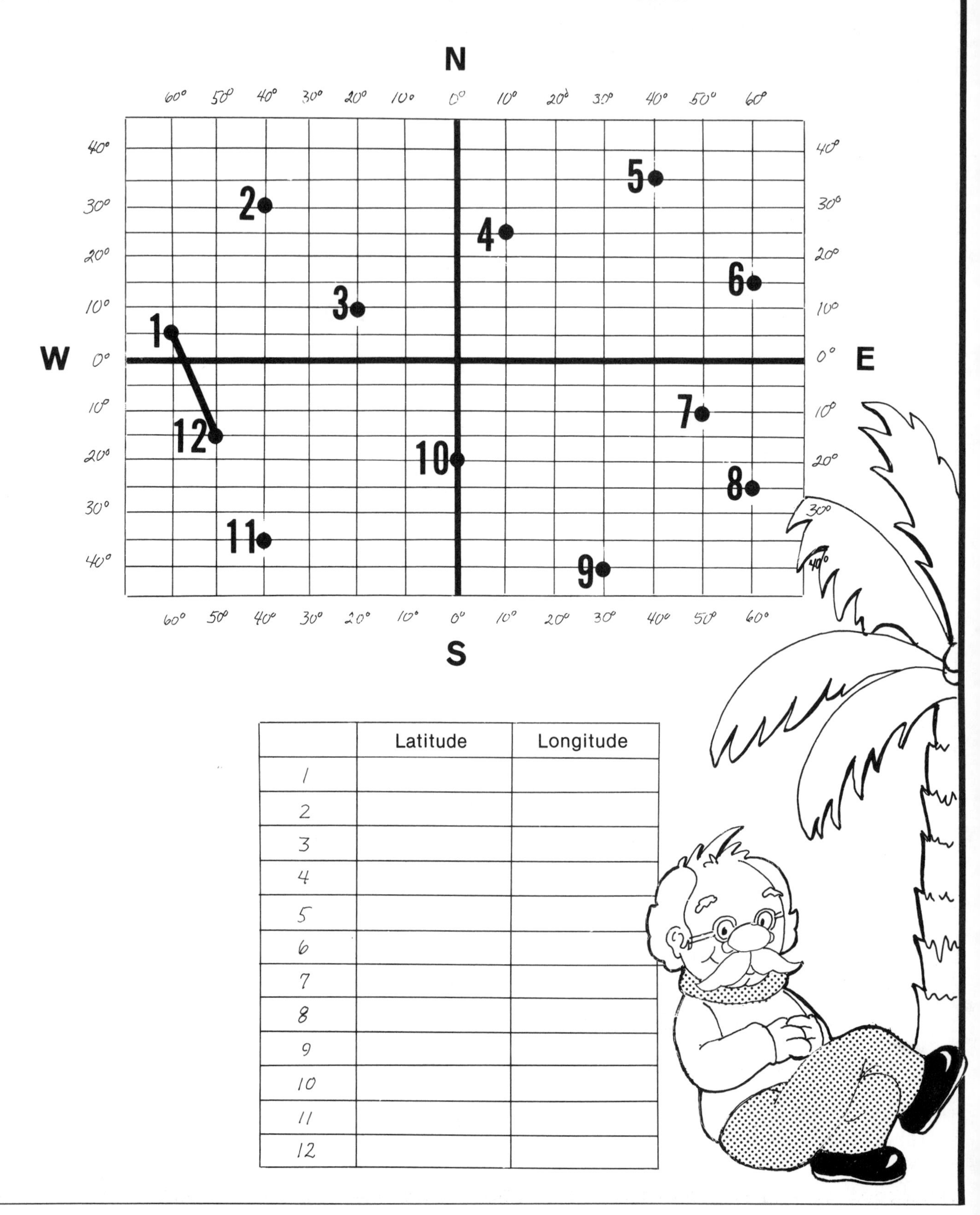

	Latitude	Longitude
1		
2		
3		
4		
5		
6		
7		
8		
9		
10		
11		
12		

NAME____________________

PBA'S MAP MAZE

Help Professor Ben Around use the road map below to guide him through his island of paradise. Find the best routes to travel by using the symbols in the legend. If you succeed in this task, you are invited to visit him for one week—all expenses paid!

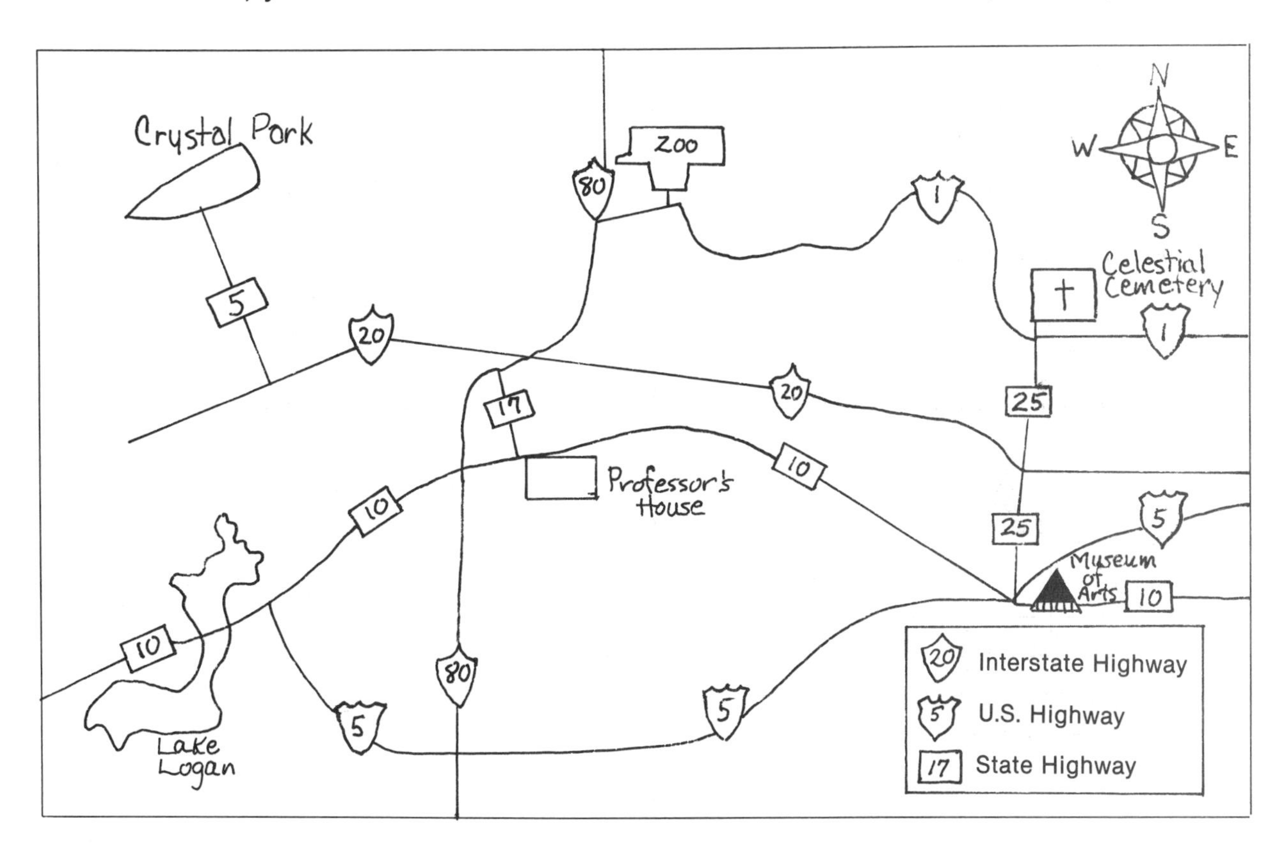

QUESTIONS:

1. What is the longest state highway? ____________________
2. Which interstate highway runs north to south? ____________________
3. What is the U.S. highway that runs between Lake Logan and Museum of Arts? ____________________
4. Which interstate highway runs east to west? ____________________
5. From the professor's house, what are the best routes to the zoo? ____________________
6. Traveling north on state highway 5 leads to what point of interest? ____________________
7. What route will take Professor Ben from his house to Lake Logan? ____________________
8. What place on the island has three different roads leading to it? ____________________

NAME____________________

LAND OF LIBERTY

The United States with its ideals of freedom and liberty continues to attract individuals from many lands. Because of its foundation on liberty, Professor Ben Around has a special name for the U.S.

To discover the name, fill in the blanks with the name of the place that is described by each clue. You may use reference books if needed. The circled letters spell the answer. Write the letters in the blanks below.

CLUES:

1. Capital of U.S. _◯_ _ _ _ _ _ _ _ _ _, _._.
2. Ocean east of U.S. _ _ _ _ ◯ _ _ _
3. River between U.S. and Mexico _ _ _ _ _ ◯ _ _ _
4. River flowing by Mount Vernon _ _ ◯ _ _ _ _ _
5. Chicago's nearest lake _ _ _ _ _ _ ◯ _ _ _ _ _ _ _
6. Largest mountain chain in U.S. _ ◯ _ _ _ _ _ _ _ _ _ _ _ _ _ _ _
7. Formation created by Colorado River _ _ _ ◯ _ _ _ _ _ _ _ _
8. Busy world port on Gulf of Mexico _ _ _ ◯ _ _ _ _ _ _ _
9. Famous Yellowstone geyser _ _ _ ◯ _ _ _ _ _ _ _ _
10. Longest river in U.S. _ ◯ _ _ _ _ _ _ _ _ _ _ _
11. Highest point in North America _ _. ◯ _ _ _ _ _ _ _
12. Vast wasteland in southwestern U.S. ◯ _ _ _ _ _ _ _ _ _ _ _ _ _
13. Easternmost state in U.S. _ _ ◯ _ _
14. The Mississippi River empties here ◯ _ _ _ _ _ _ _ _ _ _ _ _
15. Automobile captial of the world _ _ _ ◯ _ _ _
16. Hawaii's most populated island _ ◯ _ _
17. The Silver State ◯ _ _ _ _ _
18. Northernmost place in U.S. _ _ _ _ ◯ _ _ _ _ _ _
19. Largest river in New York _ _ _ ◯ _ _

ANSWER: _ _ _ _ _ _ _ _ _ _ _

_ _ _ _ _ _ _ _ _ _ _ _ _ _ _ _

BACKGROUND INFORMATION FOR THE TEACHER

Page 1: **HAWAII**

Hawaii is located in the middle of the Pacific Ocean. It is made up of islands which attract people from all over the world. Honolulu is the capital and largest city. Hawaii is the only state that does not lie on the land mass of North America.

Hawaii was formed by five volcanoes. The black sand beaches are crushed lava remains from the volcanoes. The climate of Hawaii is very extreme. There are places where frost appears every night. Mauna Loa and Mauna Kea have snow on their peaks in winter months. Hawaii is a tropical island paradise.

Page 2: **MOUNT RUSHMORE**

Located in the Black Hills of South Dakota are the faces of four American Presidents: George Washington, Thomas Jefferson, Theodore Roosevelt, and Abraham Lincoln. Drills and dynamite carved these figures out of granite cliffs under the supervision of Gutzon Borglum.

Mount Rushmore has the largest figures of any statue in the world. It rises 5,725 feet above sea level and more than 500 feet above the valley. The monument stands taller than the Great Pyramid of Egypt.

Page 3: **ALASKA**

Alaska is the largest state in the United States. However, fewer people live there than in any other state. The most populated city is Anchorage. Alaska is only fifty-one miles from Russia.

Alaska is known for its beautiful mountains and scenic sights. Its highest peak, Mount McKinley, is the highest point in North America. Alaska's chief source of wealth is found in its oil reserves at Prudhoe Bay.

Page 4: **CONTINENTAL DIVIDE**

The Continental Divide is located high in the Rocky Mountains. It separates the waters flowing into the Atlantic Ocean from those flowing into the Pacific. In the United States the Continental Divide crosses New Mexico, Colorado, Wyoming, Idaho, and Montana. It is also known as the "Great Divide."

Page 5: **GRAND CANYON**

The awesome beauty of our nation's scenic wonders unfurls as one views the Grand Canyon. Located in northern Arizona, the canyon stretches 260 miles from Utah to Nevada. It measures nearly one mile deep and in places is eighteen miles wide.

The Colorado River and its tributaries carved the Grand Canyon over a period of millions of years. It was formed through erosion and uplift. Tourists may drive around the canyon, hike across it, raft down the Colorado River, and go horseback riding and fishing. It is one of the best known tourist attractions in the world.

Page 6: **OKLAHOMA CITY, OK**

Oklahoma City is the state capital and largest populated city in Oklahoma. It is the country's third largest city in land area. The city is one of the leading centers of oil production in the United States. The state capitol rests on a major oil field. Throughout the capitol grounds and residential land, oil wells operate.

Oklahoma City is dependent on manufacturing. Industries make airplanes, food products, petroleum products, and space electronic equipment. Being rich in natural resources for both production and farming has made Oklahoma City industrially important. This city is located near the geographic center of the state.

Page 7: **YELLOWSTONE NATIONAL PARK**

Yellowstone National Park rambles through the states of Wyoming, Idaho, and Montana. It is the oldest national park in the United States. Geysers and hot springs bring excitement to all who come here. The beautiful canyons, rushing waterfalls, and brilliant lakes make this a scenic attraction.

Yellowstone's landscape was created by volcanoes and glaciers through millions of years. Below the ground surface lie about two miles of magma. Glaciers melted thousands of years ago and provided water for Yellowstone Lake. Over two million people visit this park yearly.

Page 8: **ASTRODOME**

Houston, Texas, is the home of the first football and baseball stadium with a roof—the Astrodome. Here the Houston Astros and Houston Oilers play their games. Excitement runs high as spectators crowd into this vast edifice to enjoy the gaiety of the games.

Page 9: **MAINE**

Maine is the largest New England state. It is known for the rocky shores and lighthouses lining its coast. There are scenic lakes, beautiful forests, and lofty mountains. Augusta is the capital of Maine.

Maine is a significant farming and fishing state. In the United States, Maine leads in trapping lobsters and catching sardines. It also grows potatoes and raises chickens. The major economy of Maine comes from its forests. The wood-processing industry provides many jobs for people who live there. Maine's nickname is the "Pine Tree State."

Page 10: **MT. WASHINGTON, NH**

The White Mountains of New Hampshire, located near Fabyan, are made up of eighty-six peaks. The highest one is Mount Washington which is 6,288 feet in height.

A three and one-half mile cog railroad transports sightseers to the top of this rugged mountain. It was the first cog railroad ever built in the United States. In 1934, on top of this mountain, the strongest winds ever measured reached 231 miles per hour. The scenic countryside attracts tourists from many places.

Page 11: **FOUR CORNERS MONUMENT**

The corners of Arizona, Colorado, Utah, and New Mexico come together at this location. This is the only place in the United States where this occurs. The monument at the site is surrounded by a concrete platform with the seals of all four of the states. Tourists boast that they have touched four states at one time. Four Corners Monument is a popular place to visit.

Page 12: **VERMONT**

This northeastern state is one of the smallest of the forty-nine continental states. The Green Mountains lie in the central part of the state. Skiing is a popular winter sport. Montpelier is the capital of Vermont.

Vermont is the only New England state that does not border the Atlantic Ocean. In the fall and winter months, Vermont's maple trees are tapped to produce sweet maple syrup. Probably nowhere else in the United States do fall's colors display their awesome beauty more than in Vermont's forests. Vermont is small not only in size but in population, too.

Page 13: **CAMBRIDGE, MA**

Cambridge is a major industrial and educational center of the northeastern United States. It is often called the "University City" because it is the home of so many famous universities. Harvard University is located here. It is one of the oldest educational institutions in the United States. Harvard has the world's largest university library. It contains about six million books and pamphlets.

Industrial products such as ink, fire hose, weather balloons, skates, candy, crackers, and cameras are produced here. Cambridge is located on the Charles River facing Boston.

Page 14: **BONNEVILLE SALT FLATS**

This is a large desert in northwestern Utah near Salt Lake City. It is made of extremely low, level salt beds. The salt there is as hard as cement, making for an extremely good racing surface. Many land speed records have been established at the track there.

Page 15: **COLORADO**

Colorado is one of the most scenic western states. The Rocky Mountains bisect this state from its northern to its southern border. The tourist industry is an important part of this state's industry. Many national parks are located here, such as Mesa Verde, Rocky Mountain, and Great Sand Dunes. This state is known for its winter sports and resorts.

Colorado has many famous tourist attractions. Denver is known for the U.S. Mint and the state capital. It is often called the "Mile High City" because the state capitol is built on land one mile above sea level. The world's highest suspension bridge spans the Royal Gorge at the top. Aspen is a mecca for snow-skiing enthusiasts. Colorado Springs has Pike's Peak, one of the state's highest mountains.

Page 16: **PIMLICO**

One of Maryland's most exciting events is the Preakness Stakes, an annual horse race held in May at Pimlico racetrack in Baltimore. Horse racing fans flock to Pimlico to wager money on horses and to enjoy the thrill of competition. The Preakness, the Kentucky Derby, and the Belmont Stakes comprise the "Triple Crown" of horse racing.

Baltimore is Maryland's largest city. It is located on the Patapsco River. Visitors to Baltimore can see Battle Monument, the first major war memorial in the United States, and Fort McHenry, the site where Americans fought the British in 1814.

Page 17: **ATLANTIC CITY, NJ**

Atlantic City, New Jersey, is one of the largest coastal resorts in the world. It is located on Absecon Beach, a low, sandy island separated from the mainland of New Jersey by a narrow strait and several meadows that are partly covered with water at high tide.

Atlantic City is recognized as the host city for the Miss America Pageant. The national beauty and talent competition is held annually at Convention Hall on the boardwalk. The city is also known for tourism, deep-sea fishing, and the production of glassware, paint, and furniture.

Page 18: **MUSCLE SHOALS, AL**

Muscle Shoals, Alabama, is the home of several recording studios that have attracted famous musicians throughout the United States. Located on the Tennessee River in the northwestern corner of Alabama, Muscle Shoals has been called the "Hit Recording Capital of the World." Plans are underway to construct a Country Music Hall of Fame in the area to honor America's talented recording artists.

The name "Muscle Shoals" is thought to have come from mussels that were found in the shoals of the Tennessee River. Rapids on the river created navigational hazards. In 1926 the Tennessee Valley Authority constructed Wilson Dam at Muscle Shoals to control flooding, improve navigation, and provide hydroelectric power to the area. Two nitrate plants were constructed at Muscle Shoals to manufacture munitions in an effort to prevent a dependency upon Chile and Germany as sources of nitrate.

Page 19: **CRATER OF DIAMONDS, ARKANSAS**

Visitors of the Crater of Diamonds near Murfreesboro, Arkansas, may enjoy searching for diamonds. A diamond valued at a quarter of a million dollars was found at this site. Sometimes tourists are still fortunate enough to find valuable diamonds in the diamond field.

The Crater of Diamonds field is located at the southwestern portion of Arkansas near Murfreesboro.

Page 20: **HUNTSVILLE, AL**

Huntsville, Alabama, plays a vital role in the future of America's space program. Nicknamed "Rocket City, U.S.A.," Huntsville is the site of NASA's Marshall Space Flight Center and Redstone Arsenal. In the 1950's Wernher von Braun and other scientists developed America's first large guided missiles at Redstone Arsenal. The famous *Saturn V* rocket that carried men to the moon was developed in Huntsville. Visitors to Huntsville will see many exhibits including rockets, missiles, and space equipment.

A space camp is held in Huntsville each summer for youngsters interested in pursuing a career in space exploration. Applicants from all over the United States are attracted to the program.

Huntsville is Alabama's third largest city. It is located in northern Alabama.

Page 21: **DETROIT, MICHIGAN**

The city of Detroit, Michigan, was founded in 1701 by a French colonist, Antoine de la Mothe Cadillac. Detroit lies on the southeastern border of Michigan where the Detroit River flows between the United States and Canada. It is Michigan's largest port, a gateway to international commerce.

Detroit is an important industrial city. More automobiles are manufactured in Detroit than in any other part of the U.S. It is also known for the production of machine tools, salt, paint, hardware, and business machines.

Page 22: **YOSEMITE NATIONAL PARK**

Lying about 200 miles east of San Francisco, California, in the Sierra Nevada mountains is a beautiful region known as Yosemite Valley. In 1890 Congress created Yosemite National Park for tourists to enjoy. In 1906 the park was expanded. The scenic wilderness features about 700 miles of trails, lakes, waterfalls, mountains, and wildlife. El Capitan is a famous granite mass in Yosemite that rises vertically about 3,600 feet.

Yosemite National Park is open all year round. Winter snows force the closing of the roads in the High Sierra region during the coldest months of the year.

Page 23: **LOUISIANA**

Louisiana attracts tourists from all parts of the U.S. New Orleans, famous for the Mardi Gras, is a popular city to visit. Visitors are encouraged to tour the historic French and Spanish quarters. World-famous restaurants tempt visitors to sample delicious food. Jazz musicians entertain on the streets.

The brown pelican is Louisiana's state bird. It is frequently seen along the coast. Appropriately nicknamed the "Pelican State," Louisiana is also called the "Bayou State" because of its bayous or slow-moving lake or river inlets.

The capital of Louisiana is Baton Rouge.

Page 24: **CAPE HATTERAS, NC**

Cape Hatteras is located approximately 30 miles east of North Carolina. A famous lighthouse serves as a beacon to ships at sea. So many shipwrecks occurred in the area because of the dangerous shoals and rough currents that the area was nicknamed "Graveyard of the Atlantic."

Page 25: **GREEN BAY, WISCONSIN**

Green Bay, Wisconsin, was settled in the mid-1700's. It was the first permanent settlement in the state of Wisconsin. The city of Green Bay is located at the mouth of the Fox River at the southern point of the body of water called Green Bay.

Green Bay is home of the National Football League's Green Bay Packers. It is also known for the production of paper products, cheese, and canned foods.

Page 26: **OHIO**

Ohio is an industrial state with many important products. Its central location and abundance of water and minerals have helped Ohio become a leading manufacturing state. Columbus is the capital of Ohio.

One of the Ohio's best known industries is the rubber industry centered in Akron. The industry was started in 1870 by Benjamin F. Goodrich, who became famous for automobile tires. Other industries of Ohio include meat-packing, soap, shoes, glass, and textiles.

Page 27: **GEORGIA**

Georgia, the largest state east of the Mississippi River, is known for textile production and farming. It produces peaches, peanuts, pecans, and tobacco. Its forest products are important to the state.

The capital and largest city in Georgia is Atlanta. It is a vital transportation center in the Southeast. Famous for its dogwood trees, Atlanta has been called "Dogwood City."

Page 28: **ST. LOUIS**

Rising 630 feet into the air on the bank of the Mississippi River, St. Louis's Gateway Arch is the tallest monument in the United States. Visitors can ride a small train to an observation deck where they can enjoy a spectacular view. The stainless steel monument was completed in 1965 to commemorate the westward expansion after the Louisiana Purchase of 1803. It cost $29 million to construct.

St. Louis is the busiest inland port in the United States. It is located on the western bank of the Mississippi River about ten miles south of the point where the Missouri and Mississippi Rivers meet.

Page 29: **LOOKOUT MOUNTAIN**

Visitors to Chattanooga, Tennessee, enjoy a magnificent view over 2,100 feet above the city at Lookout Mountain. Located in the southern part of the Appalachian Plateau, the scenic viewpoint attracts hundreds of tourists each year.

Page 30: **EVERGLADES NATIONAL PARK**

Situated on the southwestern tip of the Florida peninsula, Everglades National Park is home to many interesting animals including the Key deer, an endangered species. The area was established as a national park in 1947. A subtropical wilderness, Everglades National Park is the site of one of the world's largest mangrove forests.

Page 31: **LAKE WINNIPESAUKEE, NH**

Lake Winnipesaukee is New Hampshire's largest lake covering about seventy-two square miles. It has 365 islands and is located in the east-central part of the state. Long ago Indians established a village around this lake. Surrounding the valley are farms and hills. Tourists enjoy the magnificent scenery and quiet setting.

Page 32: **MESA VERDE NATIONAL PARK**

About 1400 years ago a group of Indians built high cliff dwellings of stone in the sheltered recesses of the canyon walls. These homes are located in southwestern Colorado. For years archaeologists have been trying to understand the way the Indians lived.

The federal government established the national park in 1906. In Spanish *mesa verde* means "green table." The park was named Mesa Verde National Park because of the growth of juniper and pinon pines.

Page 33: **SAN ANTONIO, TEXAS**

San Antonio is nicknamed the "Alamo City" to commemorate the renowned Battle of the Alamo fought in 1836. This historic city is the country's tenth largest city.

The San Antonio River winds its way through the heart of the city. Sightseers can tour the city in canoes or gondolas. The River Walk is lined with businesses selling a variety of goods. Salespeople dress up in native costumes as they eagerly wait on customers. What an unusual way to shop!

Page 34: **CHEYENNE, WYOMING**

This Old West town is the capital and second largest city in Wyoming. Located there is one of the world's largest intercontinental ballistic missile networks. It is a major defense center of the United States.

Frontier Days is one of the nation's best, most celebrated rodeos. It began in 1897 and has brought pleasure to a lot of cowboys. Thousands of people attend annually to enjoy the parades, Indian dances, and most of all—a splendid rodeo!

Page 35: **MASSACHUSETTS**

Massachusetts has a wealth of historic sights. Boston is the capital and largest city in the state. Many circumstances preceding the Revolutionary War occurred in Massachusetts. The Pilgrims arrived in Plymouth in 1620. Paul Revere warned the Patriots that the British were coming.

John F. Kennedy, the youngest man ever elected President of the United States, was born in this grand state.

Page 36: **PAINTED DESERT**

An inspiring sight never to be forgotten is the brilliant colors of sand in the Painted Desert, located in north-central Arizona. The desert's beauty is awesome and breathtaking. Mesas, buttes, and valleys appear throughout the desert. These were created by wind and rain cutting into shalelike volcanic ash. The Painted Desert is a scenic picture of nature in its truest form.

Page 37: **IOWA**

This midwestern state is famous for being one of the top producers of corn in the United States. It is often called the "Corn State." Iowa's capital city is Des Moines. Iowa is home to the largest popcorn processing plant in the nation and one of the largest cereal mills.

Buffalo Bill was an Indian scout and fighter. He created his own show and called it his "Wild West Show." He portrayed what life was like during the beginning days of the West.

Page 38: **DODGE CITY, KS**

Dodge City's nickname was "Cowboy Capital of the World." Located in this western town was a cemetery, Boot Hill, which was full of cowboys who were a little too lively and ended up being shot. These gunfighters were buried in their boots.

The Sante Fe Railroad came through Dodge City in 1872. Being located on the Arkansas River has helped to make Dodge City a chief commercial center of southwestern Kansas.

Page 39: **PROMONTORY, UTAH**

This site is famous because the first transcontinental railroad was completed here in 1869. The Central Pacific Railroad came eastward from Sacramento, California, while the Union Pacific built westward from Omaha, Nebraska. They were joined at Promontory by driving a golden spike in celebration of the important event. The East and West were finally linked together.

Page 40: **RUGBY, ND**

Rugby is a small town in northern North Dakota. It is not known for any important economic or historical sites. Instead, Rugby has a claim of geographical significance. It is the geographic center of North America.

Page 41: **IDAHO**

This beautiful Rocky Mountain state offers tourists white water rapids, scenic lakes, snowcapped peaks, and steep canyons. Flowing through this state is the Snake River. It is often called the "River of No Return" because the people who traveled upstream could not navigate through the churning waters. Sports enthusiasts come here to experience the excitement of the rapids.

Idaho's capital and largest city is Boise. Its number one product is potatoes. The state leads the nation in mining silver. Idaho's rugged mountains and picturesque scenes make it a popular location.

Page 42: **CARLSBAD CAVERNS**

Carlsbad Caverns is located in southwestern New Mexico. This cave contains some of the world's largest and most magnificent stalactites and stalagmites. One large room, called the Big Room, is 4,000 feet long and 625 feet wide. Water slowly ate away the rock to form Carlsbad Caverns. It took millions of years for this wonder of nature to develop.

During warm weather millions of bats fly out of the cave at dusk in search of insects.

Page 43: **LAS VEGAS, NV**

Las Vegas is the largest city in Nevada. It is a major tourist attraction in the United States. The streets of Las Vegas are lined with casinos and nightclubs. Americans spend billions of dollars each year gambling in Las Vegas. Many Hollywood and Broadway entertainers perform there to attract millions of visitors.

Page 44: **MONTANA**

Montana is the fourth largest state. Helena is the capital of Montana. This state has mountainous terrain. Located within the mountains are gold and silver. Another name for this state is the "Treasure State."

Long ago prospectors flocked to Montana searching for gold. Mining camps and towns were built overnight. Suddenly people became wealthy. Time passed and gold became difficult to find. Prospectors began to leave, and the towns became empty ghost towns.

Page 45: **NEW HAVEN, CT**

Yale University is located in New Haven, Connecticut. It is the third oldest institution of higher learning in the United States. Only Harvard and the College of William and Mary are older.

New Haven is often called the "City of Elms" because the streets are lined with elm trees. It is located on a branch of Long Island Sound about 70 miles northeast of New York City.

Page 46: **MAMMOTH CAVE**

About 100 miles south of Louisville, Kentucky, is Mammoth Cave, part of the world's largest cave system. Created in 1941, Mammoth Cave attracts almost two million visitors each year. Mammoth Cave consists mainly of limestone. Over millions of years water seeping through cracks in the limestone carved out the cave. Descending into the depths of the cave, visitors discover stalactites, stalagmites, and rocks of various shapes and colors.

Mammoth Cave is known to contain blind beetles and crayfish. Even eyeless fish live in Echo River which runs through the cave. Mammoth Cave is truly one of Kentucky's wonders.

☆ ☆

Page 47: **CRATER LAKE**
Located in Mount Mazama, an inactive volcano in Oregon's Cascade Mountains, is Crater Lake. It is the deepest lake in the United States, approaching 2,000 feet at its greatest depth. Crater Lake is six miles wide at its widest point. It has no known outlets.

In the late 1800's trout were placed in Crater Lake. In 1902 Crater Lake National Park was created. Today thousands of people enjoy its awesome beauty.

Page 48: **TENNESSEE RIVER**
The Tennessee River, the largest tributary of the Ohio River, drains approximately 41,000 square miles. Beginning at Knoxville, Tennessee, the Tennessee River flows southwestward through Tennessee and northwest to Kentucky where it joins the Ohio River.

Page 49: **MOUNT PALOMAR, CA**
The Hale telescope, one of the largest reflecting telescopes in the world, is found at Palomar Observatory in California. Its reflecting mirror is 200 inches in diameter. Scientists use the observatory to view and photograph celestial bodies.

Page 50: **COOPERSTOWN, NY**
About 80 miles west of Albany, New York, is Cooperstown, the home of the National Baseball Hall of Fame and Museum. The town was named after an early settler, William Cooper. His son, James Fenimore Cooper, was a famous writer.

Some people believe Abner Doubleday invented baseball in 1839 in Cooperstown, New York. Most historians, however, do not believe baseball truly originated with Doubleday. They prefer to accept the idea that baseball developed from an English game called rounders.

Page 51: **SAN DIEGO, CALIFORNIA**
San Diego Zoo, located in Balboa Park in San Diego, California, contains one of the largest animal collections in the world. The zoo not only allows spectators to enjoy many forms of wildlife, but it is actively involved in saving endangered species, such as the California condor and the snow leopard.

Page 52: **PHILADELPHIA, PA**
In 1731 the Library Company of Philadelphia became the first public library in the United States to offer book circulation to readers. Benjamin Franklin is credited with beginning Philadelphia's public library. Many rare books are housed in the library today.

Founded by William Penn, Philadelphia is a city full of historical landmarks. Visitors are attracted to Independence Hall, the Liberty Bell, Congress Hall, Carpenters' Hall, Betsy Ross House, many museums, and Revolutionary War sites.

Page 53: **NORTH CAROLINA**

Blackbeard was a British pirate with a colorful past. His name was Edward Teach, but he was called Blackbeard because of his long, black beard which he often braided. In the early 1700's Blackbeard instilled fear in citizens along the coasts of North Carolina and Virginia. He attacked ships and frequently came ashore.

The governor of Virginia sent a crew to bring in Blackbeard, dead or alive. He was captured off the coast of North Carolina, but he fought back with sword and pistol. He was killed in the struggle, and his head was taken to Virginia on a pole for all to see.

Page 54: **SACRAMENTO, CA**

Sacramento, California, was first founded as a town by John Sutter, Jr., in 1849. It was named for the Sacramento River. It has been the capital of California since 1854. Its founder is remembered at Sutter's Fort and Indian Museum.

While gold attracted settlers to Sacramento, its mild climate, and economic and cultural opportunities attract people today.

Page 55: **BONNEVILLE DAM, OR**

Bonneville Dam is located about 40 miles east of Portland, Oregon, on the Columbia River. It stands 197 feet high and 2,690 feet long. Because the dam elevated the water level, large ships now can travel freely up the Columbia River.

Salmon fishing is a major industry in Oregon and Washington. Salmon are born in fresh water, but they spend part of their lives in the ocean. Pacific salmon return to fresh water to spawn. To aid salmon in making the journey up the Columbia River, fish ladders resembling sloping waterfalls have been constructed at Bonneville Dam. Visitors to the dam may witness salmon making their trip upstream.

Page 56: **NIAGARA FALLS, NY**

Niagara Falls lies between the United States and Canada on the Niagara River. It is made up of two waterfalls, the American Falls in New York, and the Horseshoe Falls in Ontario. Observation points along both sides of the Niagara River offer spectators a spectacular view of the water as it plunges into a deep gorge.

Page 57: **MOUNT ROGERS**

The highest peak in Virginia is Mount Rogers at an elevation of 5,729 feet. It is located in the southern part of the Blue Ridge Mountains.

Page 58: **MOUNT RAINIER**

One of Washington's most spectacular natural attractions is Mount Rainier, the highest point in the state. It towers over 14,000 feet and has many beautiful lakes and waterfalls.

☆ ☆

Page 59: **DELAWARE**

Delaware's cultural, historical, and recreational attractions invite tourists to the state. A favorite attraction is the Henry Francis du Pont Winterthur Museum near Wilmington, housing Early American furniture. Delaware's beautiful lakes, rivers, beaches, and parks offer excellent recreational opportunities.

The capital of Delaware is Dover.

Page 60: **THE STATE OF NEW JERSEY**

New Jersey lures vacationers to its beaches, shops, and historic landmarks. Highlights of the state include Princeton University, Walt Whitman House in Camden, Atlantic City, and Edison National Site in West Orange, the home of Thomas Edison.

The capital of New Jersey is Trenton.

Page 61: **GREAT SALT LAKE**

Great Salt Lake is located in Utah. Fresh water streams flow into it. However, it is saltier than the ocean. Because the waters of the lake do not drain off, but dry up, salt is left. When the rains do not come, the water level drops, and the water becomes even saltier. Often the lake's shores appear white because of the minerals and other substances left behind.

Page 62: **MOUNT KATAHDIN, ME**

Scaling to the top of Maine's highest mountain at dawn has its special rewards—you'll be the first to see the sun rise over the United States! The highest peak is around a mile high—5,286 feet. What a glorious way to begin a day!

Page 63: **ANCHORAGE, AK**

Anchorage is located in southern Alaska, west of the Chugach Mountains. It is a place for refueling for air flights between Asia and Europe and the United States mainland and Asia.

The United States bought Alaska from Russia paying only two cents per acre. That was a bargain!

Sledding in Alaska is a major means of transportation due to the snow and ice blanketing the earth. Alaskans also enjoy dogsledding, a very popular sport.

Page 64: **CONNECTICUT**

Connecticut is the third smallest state. Hartford is the capital and one of the largest cities. Around fifty insurance companies are located in Hartford, giving it the nickname, "Insurance City."

Beautiful landscape surrounds the countryside. Lakes, forests, waterfalls, and sandy shores beckon sightseers there.

★★

Page 65: THE STATE OF RHODE ISLAND

Rhode Island is the smallest state. Providence is the capital and largest city. Its nickname is the "Ocean State."

Rhode Island has its place in history because it was the first state to declare independence from Great Britain.

Rhode Island is remembered for its famous men. Roger Williams fought for religious and political freedom. Samuel Slater built the first cotton spinning mill. Today Rhode Island is a chief producer of textiles.

Page 66: OMAHA, NE

Omaha is the largest city in Nebraska. It is one of the world's chief cattle markets and meat-packing centers. Over three million heads of livestock are slaughtered yearly. In southern Omaha, stockyards cover over ninety acres of land.

Page 67: SUN VALLEY, ID

A ski resort located in south-central Idaho attracts sports enthusiasts by the thousands. The bright sunshine and recurrent snowfalls keep this town bustling. Sleighing, skating and bowling are also enjoyed there.

Page 68: DIAMOND HEAD, HI

Diamond Head is located off Waikiki. It is one of Hawaii's best known vacation spots. It is a dead volcanic crater. Scientists are confident that it will never erupt again.

Page 69: CRYSTAL CITY, TX

On the main street of Crystal City, Texas, a statue was erected to honor "Popeye the Sailor." In this garden spot so much spinach was grown that the townspeople felt a statue was in order!

Page 70: DES MOINES, IA

Des Moines is the capital and largest city of Iowa. It is located in south-central Iowa. Iowa has the highest farm population in the United States. A significant event in Iowa is the Iowa State Fair. Many agricultural and industrial exhibits are displayed there.

Page 71: ROYAL GORGE

The Royal Gorge is located near Canon City, Colorado. The world's highest suspension bridge crosses the gorge. People may walk across the bridge, or cars may drive over the bridge. An incline railway is another way to view the canyon floor. An aerial tramway provides a sensational view.

It has taken the erosive forces of nature three million years to form the gorge. The Arkansas River has cut its path through the granite rock. Those who visit there know that they have seen one of the most remarkable wonders of nature!

Page 72: KANSAS

The capital of Kansas is Topeka. It is called the "Sunflower State," the "Wheat State," and the "Breadbasket of America." Wheat grows in every county. Kansas leads the country in producing wheat. The sunflower is the state flower of Kansas.

Dwight Eisenhower's childhood days were spent in Abilene, Kansas.

Page 73: MINNESOTA

Minnesota is sometimes called the "Bread and Butter State" because it produces wheat products, flour, and dairy products. The state is a leading producer of iron ore. The capital is St. Paul while the largest city, Minneapolis, lies on the Mississippi River. These cities are often called the "twin cities."

The longest river in the United States, the Mississippi, originates in Minnesota. Minnesota was at one time scoured by glaciers. When glaciers retreated, it caused many lakes to form. The number of lakes has been estimated at 22,000.

A huge statue salutes the legendary lumberjack, Paul Bunyan, and his ox, Babe, at Bemidji. Logging for the paper industry and lumber production is still a big part of the state's economy.

Page 74: GRAND TETONS

The Grand Tetons are located in Wyoming near Yellowstone National Park. Jackson Hole is a valley that has the Grand Tetons rising from its floor. The highest peak measures 13,770 feet.

Page 75: MARTHA'S VINEYARD

This is an island off the southeastern coast of Massachusetts. It is a famous summer resort. The island is 100 square miles in size with a population of about nine thousand people. In the summertime there are about forty thousand tourists.

Page 76: WINCHESTER, VIRGINIA

Winchester, Virginia, in the Shenandoah Valley is famous for its apple orchards. The apples are shipped to other parts of the country, processed into apple juice and vinegar, and made into delicious applesauce.

Page 77: WILLIAMSBURG, VIRGINIA

History comes alive in the historic section of Williamsburg, Virginia. Nestled between the James and York Rivers, Colonial Williamsburg covers over 170 acres. Colonial shops and buildings attract thousands of tourists who witness demonstrations of barrel making, candle dipping, and other colonial skills.

Page 78: **BARABOO, WISCONSIN**

Baraboo, Wisconsin, is the home of the Circus World Museum. The Ringling Brothers' "Greatest Show on Earth" began in Baraboo. Many early circus exhibits may be viewed in Baraboo in south-central Wisconsin.

Page 79: **BERKELEY SPRINGS, WV**

Berkeley Springs, West Virginia, is a resort area in the northeastern part of the state. Its reputation goes back to the 1700's when George Washington surveyed property for Lord Fairfax. Washington is said to have called attention to the health-granting qualities of the springs. The property was granted to the Virginia Colony by Lord Fairfax in 1776.

Page 80: **OLYMPIC NATIONAL PARK, WA**

Many tourists to Washington take advantage of the scenic campgrounds of Olympic National Park on the Olympic Peninsula near Seattle. The wilderness is home to a variety of wildlife including the world's largest herd of Roosevelt elk. President Franklin Roosevelt established Olympic National Park in 1938.

Page 81: **BROOKFIELD, IL**

Visitors delight in observing interesting animals housed at Brookfield Zoo. The animals are in a natural environment separated from viewers with moats instead of traditional bars. Brookfield is a suburb of Chicago, Illinois.

Page 82: **CUMBERLAND FALLS**

Cumberland Falls is the highest waterfall in Kentucky. It is near Corbin, Kentucky, on the Cumberland River in the southeastern part of the state. Water from the falls drops sixty-eight feet.

Page 83: **OHIO RIVER**

The Ohio River, a vital waterway in the United States, begins at Pittsburgh, Pennsylvania, and flows southward forming the borders of Ohio, Indiana, and Illinois. It forms the northern borders of West Virginia and Kentucky. It empties into the Mississippi River at Cairo, Illinois. Much industrial cargo is shipped on its waters.

Page 84: **NORFOLK, VA**

Norfolk is the largest city in Virginia. Located on the Atlantic coast, Norfolk is the home of the Norfolk Naval Base and headquarters of the Atlantic Command of the North Atlantic Treaty Organization (NATO).

Page 85: **LAKE PONTCHARTRAIN**

Lake Pontchartrain is a scenic lake in southeastern Louisiana. Covering 625 square miles, Lake Pontchartrain offers a variety of recreational opportunities. The world's longest trans-water highway, known as Lake Pontchartrain Causeway, spans the lake. It is 29.2 miles long with almost 24 miles over the water.

☆ ☆

Page 86: **MISSISSIPPI**

The state of Mississippi was named for the Mississippi River that flows along its western edge. Jackson is the capital of Mississippi. The state has rich farmlands and busy factories. It produces cotton, food products, paper, electronic supplies, and transportation equipment.

The coast of the Mississippi Gulf attracts vacationers to its beaches. Visitors may also tour old plantations and historic mansions for a feeling of life in the pre-Civil War era.

Page 87: **NEW YORK CITY**

New York City, the largest city in the United States, is a business, trade, and cultural center. It is the home of the United Nations, an organization that strives for world peace and the betterment of mankind. The United Nations Headquarters buildings overlook the East River in New York City.

Page 88: **INDIANAPOLIS, IN**

Each year during May, attention is focused on Indianapolis, Indiana, as the famous Indianapolis 500 automobile race is held. The race is held during Memorial Day weekend. Spectators flock to Indianapolis Motor Speedway to see professionals compete in the annual 500-mile race. The event is an exciting but dangerous sport that tests both automobile and driver and places Indianapolis in the international spotlight.

Indianapolis is the capital of Indiana. Located in the center of the state, it is called the "Crossroads of America." Many major highways and railroads converge there. Other attractions include the Indianapolis Zoo, Benjamin Harrison Memorial Home, and Conner Prairie Pioneer Settlement and Museum, the place where the first white people of Indianapolis settled.

Page 89: **PORTLAND, OR**

Located on the northern boundary in western Oregon, Portland is the largest city in the state. Portland has been nicknamed "City of Roses" because of the beautiful, fragrant roses that thrive in the area. Roses are featured in public as well as private gardens.

Portland is one of the busiest ports on the West Coast. It is near the point where the Columbia and Willamette Rivers join.

Page 90: **ISLE ROYALE**

Isle Royale, an island in the northwestern section of Lake Superior, is home to one of the largest herds of moose in the United States. The island is part of Isle Royale National Park, a federal game preserve in Michigan. The park consists of Isle Royale and over two hundred small islands. Many other forms of wildlife including timber wolves, squirrels, and rabbits inhabit the area.

ANSWER KEY

Introduction: LET YOUR MATH SKILLS SHINE!
CHECK YOUR WORK.

Page 1: Hawaii

C 8	A 11	N 4	I 17	W 13
Q 14	Y 12	H 15	E 20	K 24
P 10	G 28	A 11	F 6	J 30
D 22	S 16	I 17	U 18	T 44
B 36	M 46	O 34	L 26	

Page 2: Mount Rushmore

H 19	U 17	T 18	K 20	D 23
R 22	M 21	S 29	B 26	F 16
O 30	N 25	E 24	R 22	U 17
G 62	M 21	C 35	O 30	V 32

Page 3: Alaska

I 140	L 81	D 206	A 161	H 114
K 94	O 122	S 205	A 173	N 142
K 97	W 108	T 44	R 142	Y 122
A 127				

Page 4: Continental Divide

1. 75	2. 81	3. 72	4. 73	5. 62
6. 72	7. 59	8. 72	9. 73	10. 58
11. 68	12. 70	13. 62	14. 91	15. 62
16. 70	17. 59			

Page 5: Grand Canyon

A 41	D 60	C 66	N 82	A 57
N 34	G 42	N 67	R 64	O 37
Y 95				

Page 6: Oklahoma City, OK

1. 139	2. 127	3. 153	4. 79	5. 107
6. 125	7. 90	8. 178	9. 167	10. 225
11. 146	12. 159	13. 187	14. 128	15. 175
16. 123				

Page 7: Yellowstone National Park

O 74, W 90, A 109, N 83, K 91, A 67, L 85, A 81
18, 34, 53, 27, 35, 11, 29, 25
56

E 60, O 57, P 104, S 71, T 96, L 131, R 79, N 65
18, 15, 62, 29, 54, 89, 37, 23
42

E 108, I 117, Y 92, T 114, L 123, O 101, N 97
33, 42, 17, 39, 48, 26, 22
75

Page 8: Astrodome

1. 702	2. 831	3. 623	4. 994	5. 384
6. 448	7. 878	8. 641	9. 816	10. 971
11. 742	12. 789	13. 400	14. 591	15. 846

Page 9: Maine

F 2,354	E 2,396	L 1,862	K 2,759
O 984	R 1,854	H 1,690	A 2,076
W 1,655	N 1,892	S 1,272	I 1,944
T 2,000	D 2,302	M 1,393	

Page 10: Mt. Washington, NH

1. 11,001	2. 15,627	3. 15,272
4. 8,195	5. 10,577	6. 14,037
7. 10,565	8. 10,052	9. 9,022
10. 15,627	11. 14,819	12. 10,052
13. 10,052	14. 14,037	

Page 11: Four Corners Monument

N 9,983	U 7,247	R 8,505
C 49,795	M 33,691	E 136,113
T 18,354	O 7,211	F 148,436
S 24,647		

Page 12: Vermont

23,915	15,373	23,398
18,262	29,618	22,873
18,327	17,476	

Page 13: Cambridge, MA

E 505,642	A 707,442	M 174,717
I 1,668,595	R 1,155,252	B 125,162
D 1,563,414	M 1,990,847	G 168,524
A 1,047,756	C 2,289,694	

Page 14: Bonneville Salt Flats

E 531,900	N 1,146,843
R 228,212	O 7,129,068
S 1,094,955	P 117,778
A 1,324,977	B 9,934,995
L 176,847	G 101,555
T 902,205	V 215,655
I 194,508	F 84,699

Page 15: Colorado

O 12,754,083	D 17,251,238
O 9,779,117	L 11,424,036
R 15,940,777	A 16,484,040
O 18,456,233	C 8,599,000

Page 16: Pimlico

1. 516	2. 587	3. 427	4. 287
5. 460	6. 520	7. 648	8. 512
9. 511	10. 669	11. 605	12. 409
13. 725	14. 558	15. 316	

Page 17: Atlantic City, NJ

1. 7,475	2. 6,559	3. 8,829	4. 5,440
5. 6,416	6. 7,387	7. 4,556	8. 7,789
9. 5,407	10. 6,687	11. 4,829	12. 7,138
13. 5,649	14. 6,759		

Page 18: Muscle Shoals, AL

1. 6,479	2. 4,328	3. 1,168	4. 1,925
5. 2,562	6. 3,457	7. 5,369	8. 2,123
9. 1,243	10. 3,174	11. 4,783	12. 1,457
13. 2,363	14. 2,126		

Page 19: Crater of Diamonds, Arkansas

1. 254	2. 315	3. 86	4. 436	5. 322
6. 249	7. 557	8. 219	9. 256	10. 191
11. 73	12. 359	13. 321	14. 367	15. 274
16. 179	17. 313	18. 165	19. 226	20. 244
21. 335	22. 326	23. 284	24. 188	

Page 20: Huntsville, AL

1. 146 = 150	2. 288 = 290	3. 625 = 630
4. 452 = 450	5. 692 = 690	6. 489 = 490
7. 376 = 380	8. 211 = 210	9. 134 = 130
10. 267 = 270	11. 241 = 240	12. 282 = 280

Page 21: Detroit, Michigan

1. 4,526	2. 1,662	3. 2,404	4. 861
5. 3,277	6. 1,426	7. 3,564	8. 1,758
9. 3,304	10. 931	11. 1,912	12. 3,959
13. 2,594	14. 2,903	15. 3,083	

Page 22: Yosemite National Park

1. 1,962	2. 2,582	3. 442	4. 4,013
5. 3,860	6. 1,998	7. 2,106	8. 3,258
9. 2,218	10. 908	11. 3,294	12. 2,791
13. 2,202	14. 4,518	15. 1,941	16. 601
17. 2,023	18. 3,256	19. 2,869	20. 4,243

Page 23: Louisiana

1. 2,389	2. 2,891	3. 2,778	4. 2,815
5. 2,713	6. 6,578	7. 3,262	8. 1,590
9. 3,805	10. 3,281	11. 1,915	12. 3,819

Page 24: Cape Hatteras, NC

1. C 5,628	2. A 4,525	3. P 7,078
4. E 7,148	5. H 5,621	6. A 8,473
7. T 7,079	8. T 7,180	9. E 6,766
10. R 3,723	11. A 5,731	12. S 7,636
13. N 5,249	14. C 4,914	

Page 25: Green Bay, Wisconsin

1. 608	2. 476	3. 1,847
4. 1,162	5. 2,297	6. 2,176
7. 2,921	8. 2,456	9. 1,008
10. 2,622	11. 1,198	12. 1,938
13. 1,405	14. 1,922	15. 777
16. 1,829	17. 1,096	

Page 26: Ohio

1. 8,073	2. 17,577	3. 15,558
4. 15,091	5. 11,471	6. 46,879
7. 21,666	8. 7,906	9. 16,423
10. 24,617	11. 31,908	12. 10,925
13. 11,394	14. 12,165	

Page 27: Georgia

B 20,787	A 26,939	G 12,834
W 15,257	E 15,962	L 33,361
K 43,763	N 22,129	S 38,055
O 15,624	R 23,334	P 13,583
G 19,114	I 15,842	H 21,999
A 44,144		

Page 28: St. Louis

1. 8,246	2. 19,588	3. 28,445
4. 17,718	5. 10,819	6. 5,933
7. 9,436	8. 20,561	9. 17,858
10. 16,553	11. 6,604	12. 22,821
13. 9,158	14. 14,132	15. 5,843
16. 20,759		

Page 29: Lookout Mountain

1. 8,099	2. 71,209	3. 15,378
4. 13,785	5. 21,038	6. 65,239
7. 12,809	8. 19,248	9. 14,611
10. 11,971	11. 13,774	12. 52,514
13. 28,837	14. 16,721	15. 17,444

Page 30: Everglades National Park

G 842,848	O 395,622
V 216,703	I 560,908
A 558,944	P 439,924
T 245,789	E 371,531
R 424,033	N 181,908
L 285,193	S 216,936
K 167,377	D 222,152

Page 31: Lake Winnipesaukee, NH

U 82	I 198	A 69	N 260
E 96	N 72	B 360	F 224
E 132	H 85	S 140	K 476
W 375	E 166	C 192	P 171
L 189	I 343	A 208	K 135
G 256	D 117	N 738	E 108

Page 32: Mesa Verde National Park

4	8	12	16	20	24
28	32	36	40	44	48
52	56	60	64	68	72
76	80	84			

Page 33: San Antonio, Texas

S 2,475	N 3,240	A 6,615
O 5,496	S 3,992	T 5,355
N 2,723	O 2,637	A 3,296
T 5,346	N 2,625	I 6,138
A 3,024	X 3,192	E 2,784

Page 34: Cheyenne, Wyoming

Y 34,825	M 16,780	N 11,416
W 42,848	G 5,148	O 39,316
I 28,476	H 63,536	N 15,282
Y 42,714	E 9,741	E 36,799
C 15,345	N 25,024	E 32,418

Page 35: Massachusetts

M A S S A C H U S E T T S

52,452 × 6
35,625 × 9
6,237 × 4
14,987 × 3
30,318 × 7
24,536 × 6
21,538 × 9
22,353 × 5
52,389 × 7
82,725 × 8
31,894 × 8
75,676 × 9
99,452 × 9

Page 36: Painted Desert

1. 414,175	2. 471,138
3. 540,712	4. 407,916
5. 99,039	6. 116,960
7. 583,296	8. 184,238
9. 622,656	10. 253,540
11. 166,275	12. 175,062
13. 732,688	

Page 37: Iowa

29,100	1,320	46,620
1,400	780	58,380
7,120	2,610	22,410
900	8,570	820
2,520	24,960	4,900

Page 38: Dodge City, KS

1. 2,100	2. 4,536	3. 4,140
4. 3,128	5. 1,798	6. 2,436
7. 7,052	8. 754	9. 1,175
10. 2,262	11. 1,206	

Page 39: Promontory, Utah

R 11,088	T 44,416	U 42,312
O 42,048	Y 25,248	N 8,027
P 18,816	H 32,938	M 17,712
A 56,475		

Page 40: Rugby, ND

Y 206,416	O 361,257	H 484,314
U 141,453	T 748,230	R 392,291
E 302,676	N 785,862	G 222,598
D 637,056	S 508,338	B 403,354

Page 41: Idaho

D 2,958,039	E 6,194,760	W 7,476,768
M 4,926,090	H 3,484,167	A 2,386,524
O 4,067,514	S 4,779,084	N 2,506,296
I 886,121		

Page 42: Carlsbad Caverns

B	157,872	A	166,770	S	647,797
V	412,438	N	604,996	E	457,312
L	128,243	R	542,290	C	83,692
S	146,888	A	87,576	C	249,275
D	172,386	R	98,754	A	332,640

Least to greatest:

C	83,692	A	87,576	R	98,754
L	128,243	S	146,888	B	157,872
A	166,770	D	172,386	C	249,275
A	332,640	V	412,438	E	457,312
R	542,290	N	604,996	S	647,797

Page 43: Las Vegas, NV

A	1,255,105	G	1,150,892	V	2,146,875
S	2,436,786	A	468,531	N	5,597,408
S	3,186,882	V	1,262,472	L	1,343,106
E	2,570,344				

Page 44: Montana

A $7^3 = 343$
N $2^6 = 64$
O $4^6 = 4,096$
A $6^4 = 1,296$
N $10^6 = 1,000,000$
M $5^4 = 625$
T $10^3 = 1,000$

Page 45: New Haven, CT

H	25,292,251	E	18,792,690	T	3,233,164
E	22,131,347	N	11,313,820	V	41,124,240
C	25,666,055	W	6,380,850	A	20,089,342
N	36,197,402				

Page 46: Mammoth Cave

1. 7 2. 8 3. 56 4. 3 5. 6
6. 28 7. 72 8. 5 9. 11 10. 49
11. 40 12. 9 13. 4 14. 24 15. 2
16. 10 17. 15 18. 12 19. 36 20. 1
21. 13 22. 63 23. 16 24. 48

Page 47: Crater Lake

B. 3 R4	C. 5 R5	O. 5 R4	R. 5 R1
A. 6 R1	T. 6 R2	S. 9 R1	E. 2 R6
R. 6 R3	Y. 4 R3	L. 7 R1	W. 5 R6
A. 5 R3	P. 6 R5	K. 8 R1	E. 5 R2

Page 48: Tennessee River

1. 17 R4 2. 18 3. 11 R6 4. 12 R4
5. 17 R2 6. 19 R1 7. 15 R2 8. 14 R3
9. 16 10. 11 R5 11. 21 R1 12. 17
13. 12 R2 14. 15 R1 15. 12

Page 49: Mount Palomar, CA

1. 172 2. 201 3. 103 4. 141
5. 125 6. 102 7. 116 8. 134
9. 147 10. 175 11. 115 12. 304
13. 182 14. 121

Page 50: Cooperstown, NY

1. 203 2. 106 3. 504 4. 602
5. 109 6. 404 7. 707 8. 605
9. 703 10. 401 11. 508 12. 603
13. 802 14. 706 15. 507 16. 601

Page 51: San Diego, California

1. 31 2. 30 3. 62 4. 71
5. 80 6. 52 7. 91 8. 50
9. 21 10. 120 11. 201 12. 105
13. 103 14. 132 15. 104 16. 108
17. 93 18. 109

Page 52: Philadelphia, PA

1. 122 R2 2. 234 R2 3. 58 R4 4. 157 R3
5. 43 R3 6. 134 R3 7. 148 R2 8. 89 R2
9. 51 R3 10. 38 R6 11. 133 R6 12. 103 R1
13. 58 R8 14. 72 R3

Page 53: North Carolina

1. 363 R20 2. 239 R30 3. 77 R29 4. 111 R44
5. 123 R4 6. 97 R9 7. 79 R7 8. 28 R26
9. 110 R23 10. 68 R19 11. 52 R43 12. 156 R2
13. 86 R52

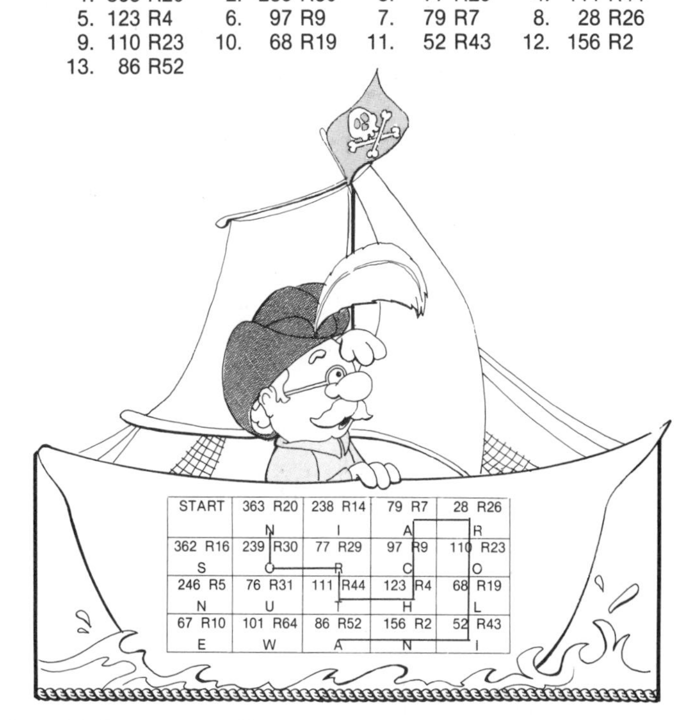

Page 54: Sacramento, CA

S	A	C	R	A	M	E	N	T	O,	C	A
14R6	10R20	7R13	13R4	11R14	5R5	7R14	6R16	6R24	8R16	18R12	12R6

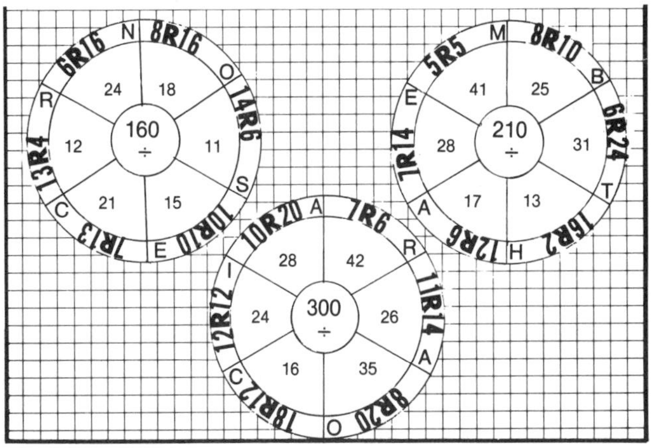

Page 55: Bonneville Dam, OR

1. 5 2. 4 3. 3 4. 11
5. 8 6. 9 7. 7 8. 16
9. 2 10. 12 11. 6 12. 19
13. 18 14. 15 15. 22

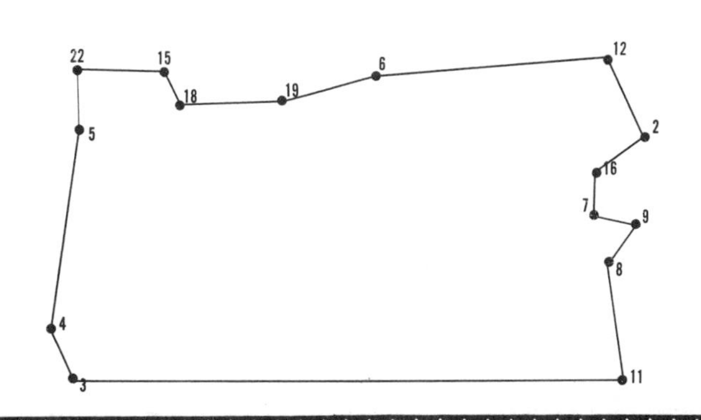

Page 56: Niagara Falls, NY

R 17 L 23 F 98 G 18
Y 26 A 42 N 63 S 71
T 85 I 65 B 68 O 49
A 95 L 69 E 62 A 24
C 28 A 37 H 51 N 38

Page 57: Mount Rogers

M 412 T 369 C 136 O 503
O 209 U 103 K 247 N 297
T 256 S 106 R 148 B 247
A 123 O 234 T 106 G 101
E 158 Y 147 R 192 S 114

Page 58: Mount Rainier

N 32 I 58 O 61 R 78 M 93 E 25
I 43 A 50 R 83 N 27 U 19 T 34

Page 59: Delaware

1. 36 R1 2. 16 R20 3. 21 R3 4. 12 R2
5. 24 R4 6. 17 R2 7. 19 R6 8. 11 R7
9. 56 R16 10. 16 R24 11. 14 R52 12. 15 R16

Page 60: The State of New Jersey

1. 296 2. 276 3. 306 4. 541
5. 695 6. 741 7. 592 8. 325
9. 195 10. 184 11. 680 12. 203
13. 453 14. 486 15. 324 16. 524
17. 318 18. 429 19. 286

Page 61: Great Salt Lake

1. G 2. R 3. E 4. A
5. T 6. S 7. A 8. L
9. T 10. L 11. A 12. K
13. E

Page 62: Mount Katahdin, ME

7 4 7 18
1. M 2. O 3. U 4. N
25 12 24 6
5. T 6. K 7. A 8. T
24 28 25 27
9. A 10. H 11. D 12. I
12 36 44
13. N 14. M 15. E

Page 63: Anchorage, AK

3C 12L 20T 18A
5N 10R 9A 8H
2P 6K 4O 16A
24M 35E 15G

Page 64: Connecticut

2 1/6 4 | 5 7/10 6 3/4
3 2 1/2 | 5 1/4 4 5/8
2 3/7 1 1/3 | 2 1/3 4 4/5
1 1/12 3 1/6 | 1 2/9 9

Page 65: The State of Rhode Island

1. 7/11 2. 8/9 3. 2/3 4. 1/5 5. 3/7
6. 2/5 7. 3/40 8. 4/11 9. 1/12 10. 1/2
11. 3/8 12. 3/4 13. 4/7 14. 7/9 15. 4/5
16. 6/7 17. 5/9 18. 3/5 19. 1/4 20. 1/6
21. 5/6

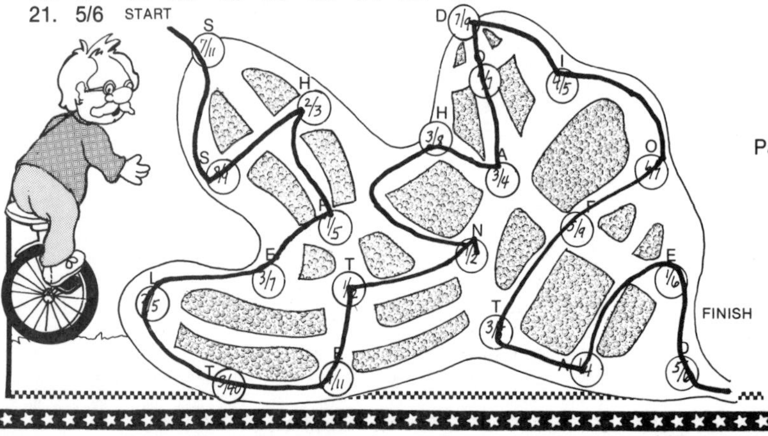

START

FINISH

Page 66: Omaha, NE

F 1 1/6 O 1/2 I 1 1/8 M 5/12
D 1 1/8 A 6/7 T 1 2/5 H 8/9
O 1 1/4 N 1 3/11 A 5/8 W 1 1/2
L 2 N 13/18 S 1 1/14 E 2/3

Page 67: Sun Valley, ID

M 1/6 A 1/14 F 1/5 L 1/2
S 3/5 V 5/9 D 3/4 S 2/9
Y 1/8 P 1/12 E 1/7 O 3/13
W 1/10 U 2/5 T 1/3 L 3/17
R 5/14 N 2/25 C 7/11 I 7/18

Page 68: Diamond Head, HI

D 5/8 F 1 1/2 C 1 11/36 I 5/12
L 1 1/12 A 13/20 L 1 1/24 M 2/3
I 1 3/8 O 9/10 U 1 2/9 N 31/40
D 7/12 R 1 7/24 H 4/5 I 1 4/15
E 5/9 A 5/6 G 1 1/6 D 19/20
H 1/2 I 13/15

Page 69: Crystal City, TX

S 1/12 T 19/40 A 5/12 W 7/27
X 3/4 P 2/9 C 13/24 L 2/5
Y 13/18 T 2/15 B 3/5 I 7/16
U 3/8 H 1/2 R 7/24 F 5/16
G 3/10 Y 1/6 O 31/63 T 11/36
C 1/5

Page 70: Des Moines, IA

D 20 1/4 E 12 1/6 S 7 2/5 M 16
O 4 5/11 I 4 1/4 N 2 1/2 E 10
S 9 4/9 I 5 5/12 A 4 4/7

Page 71: Royal Gorge

O 3 13/20 A 16 1/6 G 5 1/10
R 17 13/18 E 8 3/8 O 41 1/3
Y 11 1/10 R 30 17/21 L 4 7/30
G 20 1/2

Page 72: Kansas

1. 1/8 2. 3/16 3. 3/14 4. 1/4
5. 1/24 6. 2/9 7. 2/15 8. 1/16
9. 1/3 10. 1/9 11. 5/12 12. 1

Page 73: Minnesota

1. 3 8/9 2. 9 7/20 3. 14 1/6
4. 20 7/12 5. 14 14/15 6. 3 2/9
7. 18 3/8 8. 4 11/16 9. 77 19/28

Page 74: Grand Tetons

1. 1 1/2 2. 8/15 3. 2 2/5 4. 1 5/16
5. 6 6/7 6. 1 4/5 7. 1/2 8. 6
9. 15/28 10. 5/6 11. 3/5 12. 2/3
13. 1 3/25 14. 5/12 15. 3 1/3 16. 2
17. 9/16 18. 4

Page 75: Martha's Vineyard

R 2 17/20 E 1 5/14 T 3 4/5 V 21/25
M 2 4/63 I 2 16/25 A 13/22 N 31/40
D 3 15/32 Y 45/92 H 3/8 S 19/46
A 27/56 R 2 17/64 A 4/5

Page 76: Winchester, Virginia

1. 3.7 2. 1.8 3. 4.06 4. 2.02
5. 4.006 6. 2.002 7. 2.2 8. 5.09
9. 3.007 10. 6.004 11. 4.6 12. 1.008
13. 6.04 14. 6.4 15. 5.9 16. 5.009
17. 1.08 18. 3.07

Page 77: Williamsburg, Virginia

1. W 3.50 2. I 1.3 3. L .46 4. L 2.99
5. I .33 6. A 5.808 7. M 1.20 8. S 3.76
9. B 7.25 10. U .47 11. R .25 12. G .610
13. V 1.25 14. I 1.40 15. R 2.90 16. G .350
17. I 1.07 18. N 3.63 19. I 1.78 20. A 6.50

Page 78: Baraboo, Wisconsin

1. B 5.2	2. A 3.5	3. R 2.5	4. A 6.2
5. B 5.4	6. O 8.7	7. O 11.5	8. W 7.8
9. I 5.0	10. S 2.4	11. C 3.5	12. O 9.1
13. N 5.6	14. S 7.5	15. I 1.5	16. N 8.9

Page 79: Berkeley Springs, WV

1. 5	2. 3	3. 11	4. 4
5. 13	6. 10	7. 8	8. 7
9. 6	10. 17	11. 14	12. 9
13. 12	14. 15	15. 16	16. 18
17. 19			

Page 80: Olympic National Park, WA

	O	L	Y	E	L	S	M	B
1.2	1.5	1.8	2.0	2.3	2.5	3.6	3.9	4.1
S	N	Y	L	I	N	A	T	L
1.7	2.2	2.1	2.3	2.6	2.9	3.2	4.3	4.6
E	Q	M	P	T	S	R	K	A
2.0	2.3	2.4	2.7	3.1	3.4	3.7	4.0	4.9
O	U	C	I	J	O	N	A	C
2.9	2.6	3.3	3.0	4.0	4.8	5.1	5.4	5.1
E	R	N	A	T	I	B	L	T
5.8	5.5	3.6	3.9	4.2	4.5	5.0	5.7	5.4
D	K	R	E	W	B	N	P	E
6.1	5.2	4.9	4.0	5.9	5.6	5.3	6.0	6.1
W	O	A	C	R	T	R	A	I
6.4	6.7	4.6	4.3	6.1	6.4	6.6	6.3	6.5
D	O	P	A	C	M	K	W	A
6.2	7.0	4.2	4.5	6.2	6.7	6.9	7.2	7.5

Page 81: Brookfield, IL

1. 21.1	2. 27.3	3. 17.8	4. 21.9
5. 22.6	6. 22.7	7. 12.2	8. 14.5
9. 25.8	10. 16.8	11. 19.9	12. 31.7
13. 13.1	14. 18.9	15. 21.8	16. 25.4
17. 26.2	18. 32.5	19. 30.8	20. 20.5

Page 82: Cumberland Falls

1. 16.3	2. 25.8	3. 24.4	4. 19.9
5. 33.78	6. 43.79	7. 25.32	8. 40.32
9. 49.56	10. 44.62	11. 36.80	12. 20.72
13. 59.82	14. 56.26	15. 54.68	

Page 83: Ohio River

1. 178.57	2. 357.29	3. 504.454
4. 332.5	5. 253.26	6. 787.84
7. 265.869	8. 152.368	9. 361.769
10. 281.111	11. 43.527	

Page 84: Norfolk, VA

1. 21.0	2. 24.0	3. 18.6	4. 15.6
5. .30	6. 1.44	7. 3.60	8. 1.23
9. 4.25	10. 5.44	11. 15.75	12. 26.88

Page 85: Lake Pontchartrain

1. 59.5	2. 55.971	3. 752.4
4. 720.174	5. 1,166.2	6. 3,207.96
7. 10,643.58	8. 1,151.018	9. 1,019.35
10. 642.642	11. 290.852	12. 4,155.36
13. 6,968.5	14. 15,516.86	15. 59.3217
16. 203.2044	17. 1,466.994	

Page 86: Mississippi

1. 20.7788	2. 5.5552	3. 13.2732
4. 9.2085	5. 209.304	6. 167.128
7. 376.809	8. 352.684	9. 3,653.66
10. 7,780.47	11. 13,599.8	12. 3,614.22
13. 7,154.91	14. 9,818.82	15. 8,618.4
16. 15,883.28	17. 1,850.752	18. 4,493.944
19. 1,882.416	20. 849.312	

Page 87: New York City

C 1.82	W 1.89	Y 1.69
N 1.45	Y 1.82	K 1.33
A .52	I 1.35	B .71
L .60	D .41	O 1.18
E 1.22	R 1.06	T 2.12
V .90	G .87	

Page 88: Indianapolis, IN

D 14.8	A 20.4	N 21.8
N 12.25	I 0.2	I 15.4
L 125	P 103	A 102
I 130	O 112	N 515
S 140	I 177.5	

Least to greatest:

I 0.2	N 12.25	D 14.8
I 15.4	A 20.4	N 21.8
A 102	P 103	O 112
L 125	I 130	S 140
I 177.5	N 515	

Page 89: Portland, OR

1. 3.6	2. 6.3	3. 4.2
4. 4.9	5. 9.3	6. 5.2
7. 3.1	8. 3.8	9. 3.9
10. 5.7	11. 7.3	12. 6.7

Page 90: Isle Royale

1. 55.9	2. 86.8	3. 701.1
4. 33.86	5. 1.35	6. 21.28
7. 35.75	8. 320.151	9. .91
10. 6.32	11. 6.7	12. 105

Page 92:

1. Kentucky	2. Arkansas	3. Texas
4. Tennessee	5. Arizona	6. New York
7. Alabama	8. New Mexico	9. Indiana
10. Idaho		

Page 94:

1. Alabama	AL	Montgomery
2. Alaska	AK	Juneau
3. Arizona	AZ	Phoenix
4. Arkansas	AR	Little Rock
5. California	CA	Sacramento
6. Colorado	CO	Denver
7. Connecticut	CT	Hartford
8. Delaware	DE	Dover
9. Florida	FL	Tallahassee
10. Georgia	GA	Atlanta
11. Hawaii	HI	Honolulu
12. Idaho	ID	Boise
13. Illinois	IL	Springfield
14. Indiana	IN	Indianapolis
15. Iowa	IA	Des Moines
16. Kansas	KS	Topeka
17. Kentucky	KY	Frankfort
18. Louisiana	LA	Baton Rouge
19. Maine	ME	Augusta
20. Maryland	MD	Annapolis
21. Massachusetts	MA	Boston
22. Michigan	MI	Lansing
23. Minnesota	MN	St. Paul
24. Mississippi	MS	Jackson
25. Missouri	MO	Jefferson City
26. Montana	MT	Helena
27. Nebraska	NE	Lincoln

28. Nevada	NV	Carson City
29. New Hampshire	NH	Concord
30. New Jersey	NJ	Trenton
31. New Mexico	NM	Santa Fe
32. New York	NY	Albany
33. North Carolina	NC	Raleigh
34. North Dakota	ND	Bismarck
35. Ohio	OH	Columbus
36. Oklahoma	OK	Oklahoma City
37. Oregon	OR	Salem
38. Pennsylvania	PA	Harrisburg
39. Rhode Island	RI	Providence
40. South Carolina	SC	Columbia
41. South Dakota	SD	Pierre
42. Tennessee	TN	Nashville
43. Texas	TX	Austin
44. Utah	UT	Salt Lake City
45. Vermont	VT	Montpelier
46. Virginia	VA	Richmond
47. Washington	WA	Olympia
48. West Virginia	WV	Charleston
49. Wisconsin	WI	Madison
50. Wyoming	WY	Cheyenne

Page 95:

S	C	H	A	N	T	L	O	C	B	E	T	R	U	E	N	A	V	S	L	D
F	I	J	L	A	N	D	F	O	R	M	J	K	S	L	F	M	O	P	O	T
C	R	T	W	E	Y	S	X	R	I	V	E	R	E	Z	B	S	Y	U	N	L
H	E	Z	I	S	T	H	M	U	S	B	Z	J	A	E	Z	S	O	R	G	A
Y	S	T	P	L	A	T	E	A	U	L	G	N	D	U	K	T	I	Y	I	K
F	E	R	Z	S	B	H	I	Y	M	A	Q	U	B	H	L	R	E	D	T	S
A	R	T	S	R	E	N	A	J	F	K	T	B	L	A	N	A	S	E	U	W
J	V	B	D	M	W	B	S	K	X	I	W	S	Y	F	W	I	N	L	D	R
B	O	T	V	A	L	L	E	Y	T	U	S	I	T	A	Z	T	R	T	E	O
E	I	R	S	Y	T	Q	U	A	C	A	N	Y	O	N	F	J	O	A	B	S
X	R	I	A	D	J	K	L	T	X	W	S	F	P	L	A	I	N	W	Y	P
H	E	B	Y	W	G	B	A	S	I	N	H	Z	J	B	O	B	R	P	S	E
T	S	U	A	V	M	O	U	N	T	A	I	N	A	D	G	A	Z	R	A	N
A	Z	T	Q	U	A	N	X	F	R	Q	V	S	N	W	S	D	B	A	E	I
C	H	A	G	I	N	L	E	T	H	S	Y	A	K	I	O	J	L	I	Y	N
I	J	R	N	K	R	W	E	I	O	T	L	T	Z	S	U	E	A	R	O	S
B	S	Y	E	A	S	N	O	X	F	S	G	D	A	Y	R	G	K	I	S	U
M	E	R	I	D	I	A	N	W	I	Z	H	Q	U	K	C	O	E	E	F	L
H	O	N	S	K	V	E	Z	L	M	Y	R	N	P	O	E	Y	N	X	A	A

Page 96:

1. D	2. F	3. I
4. J	5. L	6. B
7. O	8. M	9. P
10. Q	11. G	12. C
13. H	14. R	15. A
16. N	17. K	18. E

Page 99:

Gulf of Fear to Ship Island	2½″ = 62.5 mi.
Ship Island to Peninsula	2″ = 50 mi.
Peninsula to Delta	1½″ = 37.5 mi.
Delta to White River	3½″ = 87.5 mi.
White River to Pirate's Lake	2″ = 50 mi.
Pirate's Lake to Blue Mountain	3″ = 75 mi.
Blue Mountain to Indian Plateau	2½″ = 62.5 mi.
Indian Plateau to Inlet	½″ = 12.5 mi.

Total miles = 437.5
Total days = 11

Page 100:

1. 5°N 60°W	2. 30°N 40°W
3. 10°N 20°W	4. 25°N 10°E
5. 35°N 40°E	6. 15°N 60°E
7. 10°S 50°E	8. 25°S 60°E
9. 40°S 30°E	10. 20°S 0°
11. 35°S 40°W	12. 15°S 50°W

Page 101:

1. 10
2. 80
3. 5
4. 20
5. State highway 17, Interstate highway 80, and U.S. highway 1
6. Crystal Park
7. State highway 10
8. Museum of Arts

Page 102: A Nation of Immigrants

1. Washington, D.C.
2. Atlantic
3. Rio Grande
4. Potomac
5. Lake Michigan
6. Rocky Mountains
7. Grand Canyon
8. New Orleans
9. Old Faithful
10. Mississippi
11. Mt. McKinley
12. Mojave Desert
13. Maine
14. Gulf of Mexico
15. Detroit
16. Oahu
17. Nevada
18. Point Barrow
19. Hudson